SAP Excellence

Series editors:

Prof. Dr. Dr. h.c. mult. Peter Mertens,
University of Erlangen-Nuremberg

Dr. Peter Zencke,
SAP AG, Walldorf

Springer
Berlin
Heidelberg
New York
Barcelona
Hong Kong
London
Milan
Paris
Tokyo

Gerhard Knolmayer · Peter Mertens
Alexander Zeier

Supply Chain Management Based on SAP Systems

Order Management in Manufacturing Companies

With 81 Figures
and 19 Tables

Springer

Prof. Dr. Gerhard Knolmayer
University of Bern
Institute of Information Systems
Information Engineering Group
Engehaldenstrasse 8
CH-3012 Bern
Switzerland

Prof. Dr. Dr. h.c. mult. Peter Mertens
University of Erlangen-Nuremberg
Department of Information Systems I
Lange Gasse 20
D-90403 Nuremberg
Germany

Dipl.-Kfm. Alexander Zeier
University of Erlangen-Nuremberg
Bavarian Information Systems Research Network (FORWIN)
Äußerer Laufer Platz 13–15
D-90403 Nuremberg
Germany

ISBN 3-540-66952-3 Springer-Verlag Berlin Heidelberg New York

Library of Congress Cataloging-in-Publication Data applied for
Die Deutsche Bibliothek – CIP-Einheitsaufnahme
Knolmayer, Gerhard: Supply chain management based on SAP systems: order management in manufacturing companies / Gerhard Knolmayer; Peter Mertens; Alexander Zeier. – Berlin; Heidelberg; New York; Barcelona; Hong Kong; London; Milan; Paris; Tokyo: Springer, 2002
(SAP excellence)
ISBN 3-540-66952-3

Springer-Verlag Berlin Heidelberg New York
a member of Springer Science+Business Media

http://www.springeronline.com

Hardcover Design: Erich Kirchner, Heidelberg

SPIN 11369127 42/3111-5 4 – Printed on acid-free paper

Preface

In recent years the concept of Supply Chain Management (SCM) has attracted a great deal of attention. The idea of lean management, which is aimed at preventing waste within companies, is extended to the complete supply network. Most publications on SCM stress marketing, logistics or organizational issues, whereas the role of information technology (IT) within SCM is not always accorded the attention it deserves.

A seamless collaboration of SCM software with Enterprise Resource Planning (ERP) systems is mandatory. All main vendors of ERP systems therefore provide SCM systems that are similar in many aspects. Because of its leading position in the ERP market, the mySAP SCM initiative of SAP and its SCM solutions are especially important in the realization of modern information systems (IS). Therefore this book combines the theoretical reasoning behind business administration, management concepts, and SCM-relevant methods of Operations Research with descriptions of the procedures and processes implemented in the SAP software packages R/3 and APO (Advanced Planner and Optimizer). The book focuses on intra- and inter-company order management systems. We outline the actual discussions on the state of the art support provided by IT systems for the various business functions and discuss the interfaces and the cooperative solutions between these different functions from a process viewpoint.

The English version of this book was finished in May 2001. The book refers to SAP R/3 Release 4.6 and to SAP APO Release 3.0. It has been substantially enlarged and updated compared with the German edition, which appeared in early 2000. In particular, in the discussion of the concepts of logistics management we draw more intensively on the American literature and cases.

In an area that is developing as quickly and in as many ways as SCM a book is not an appropriate medium for a detailed description of the most recent status. Therefore, we consider the aspects of IS development and integration at a fair level of abstraction, being convinced that interested readers will be able to find more detailed information for themselves, for example in white papers, presentation slides and vendors' manuals. A great deal of detailed information can also be obtained from the web presentations of vendors, consultancies, and universities. A summary of the most important SCM-related web-sites can be found in the final chapter of this book. With respect to the international readership of this book, all dates are given according to the ISO 8601 standard, i.e. in the YYYY-MM-DD notation.

This book appears in the series "SAP Excellence" in which additionally the following titles are available:

- Appelrath, Hans-Jürgen; Ritter, Jörg: SAP R/3 Implementation: Methods and Tools
- Becker, Jörg; Uhr, Wolfgang; Vering, Oliver: Retail Information Systems Based on SAP Products
- Buxmann, Peter; König, Wolfgang: Inter-organiziational Cooperation with SAP Systems - Perspectives on Logistics and Service Management

The books in this series are unique, partly because they have been written in close cooperation with experts at SAP. The authors thus had the privilege of taking part in frequent discussions giving them access to information not usually available to writers who are not employed by SAP. On the other hand, our position in the scientific community allows us to provide independent views on the various issues. Special thanks for support in the work of compilation go to Dagmar Fischer-Neeb, Anja Kutscher and to Christian Baeck, Dr. Heinrich Braun, Albrecht Diener, Dr. Georg Doll, Ulrich Eisert, Claus Grünewald, Prof. Dr. Claus Heinrich, Dr. Martin Kühn, Claudius Link, Bernhard Lokowandt, Claus Neugebauer, Andreas Pfadenhauer, Jörg Rosbach, Ralph Schneider, Martin Wallner, Andreas Wojtyczka, and Wilhelm Zwerger. Our main contact at SAP was Martin Edelmann, to whom we are indebted for all the know-how he has shared with us and for his helpful suggestions. SAP staff members were always accessible for (sometimes critical) discussions. Many thanks also go to the managers of four companies who provided help and support in formulating the case studies (cf. section 5): Dave Agey of Colgate-Palmolive, Alfred Haas and Michael Mader of fischerwerke, Volker Heilmann and Manfred Krüll of the Salzgitter Group, and Johann Thalbauer and Dr. Klaus Godzik of Wacker Siltronic.

Several members of our own staff have been closely involved in various aspects of the work on this book; their enthusiasm and support is highly appreciated. In particular, Monika Bartholdi, Andreas Billmeyer, Dr. Heide Brücher, Gerald Butterwegge, Marc Carena, Irina Depperschmidt, Rainer Endl, Jörg-Michael Friedrich, Manuel Haag, Alexandra Heeb, Marc Heissenbüttel, Mihael Höhl, Martin Hollaus, Oliver Klaus, Diego Liechti, Dr. Marco Meier, Dr. Martin Meyer, Marc-André Mittermayer, Corinne Montadon, Dr. Thomas Myrach, Marcel Pfahrer, Michael Röthlin, Waltraud Rück, Marius Stadtherr, Reimer Studt, Dr. Momme Stürken, Ulf J. Timm, and Konrad Walser each deserve special mention for their support in compilation of the German and/or the English edition. Valuable hints were given by Prof. Dr. Georg Disterer, of the University of Applied Sciences in Hanover. Many improvements to our "Germenglish" formulations were suggested by Janet Dodsworth. Many thanks go to them, and indeed to all other persons who gave us suggestions and ideas derived from both theoretical and practical points of view.

Our main hope is that the book will benefit various reader communities in stimulating the development of competitive organizational processes and IS. It is directed at managers responsible for IS or for logistics groups and provides an overview of the recent developments in the intra- and inter-company issues relating to materials, information, and financial flows. We have used the material of this book in our lectures and seminars at the Universities of Bern, Erlangen-Nuremberg, and the SCM Executive program at the University of California at Los Angeles; we also found it highly suitable for practically oriented courses on the design of information systems.

Bern and Nuremberg,
June 30, 2001 Gerhard Knolmayer Peter Mertens Alexander Zeier

Table of Contents

"The Supply Chain is, in fact, a gold mine for improving operations and profitability."
(Hicks 1997, p. 29)

1 Fundamentals of Supply Chain Management

In conjunction with extensive reorganizations of business structures and processes, the concept of "Supply Chain Management" (SCM) is rapidly gaining importance. Many companies (with different degrees of success) have realized so-called reengineering projects. They have changed structures and processes, and also reduced the degree of vertical integration, obtaining ever more products and services from external suppliers. With concepts such as virtual companies, extended enterprises, strategic alliances, and company networks, the legal and business limits of companies become blurred. Consequently, the coordination of business processes beyond the elementary organization units gains particularly in importance. Whereas "lean management" tries to counter various forms of waste within a company, SCM aims at avoiding this all along the value chain. This means there are close connections between SCM and lean management (cf. Taylor/Brunt 2001).

1.1 Order Management, Logistics, and Supply Chain Management

Logistics is usually defined to be the planning and control of material and information flows. Logistical considerations are typically based on the perspective of a single company; this can be seen by the differentiation still widely made between procurement, production, distribution, and disposal logistics (cf. Pfohl 2000, pp. 17). "Order management" is the main topic of this book, and we concentrate particularly on planning, scheduling, controlling, and monitoring this process.

The more companies are involved in producing services and products, the better they can concentrate on their core competencies. This, however, also increases the number of interfaces between them. Consequently, overall planning, scheduling, controlling, and monitoring activities of inter-company processes gain importance.

In the following, we differentiate between intra- and inter-company forms of Supply Chains (SC).

Four main uses of the term "Supply Chain Management" can be distinguished (Harland 1996, p. S64): Managing

- the internal SC, which integrates business functions involved in the flow of materials and information from the inbound to the outbound end of a business,
- dyadic, i.e., two-party relationships with direct partnerships,
- a chain of businesses including
  - a supplier, a supplier's supplier, and so on, and/or
  - a customer, a customer's customer, and so on, and
- a network of interconnected businesses involved in the ultimate provision of product and service packages required by end customers.

In three of the four definitions above the SCM concept goes beyond the single company's consideration of its material and information flows: "The integration of all key business processes across the supply chain is what we are calling supply chain management" (Cooper et al. 1997).

SCM focuses as well on operational as on strategic aspects, whereas logistics concentrates on operative issues (cf. Oliver/Webber 1982; Ross 1999, pp. 2). Differences between SCM and logistics are also emphasized in Cooper et al. (1997). In probably the first publication to use the term "supply chain management", Oliver and Webber (1982) argue that SCM differs significantly from traditional forms of materials and manufacturing control in four respects:

1. The SC should be viewed as a single entity rather than relegating fragmented responsibility for various segments in the SC to functional areas
2. SCM stresses the need for strategic decision making
3. SCM provides a different perspective on inventories, which are used as the balancing mechanism of last, not first, resort
4. SCM requires a new approach to IS based on integration instead of interfaces.

In 1998, the Council of Logistics Management modified its definition of logistics to emphasize that logistics is a subset of SCM. The revised definition is: "Logistics is that part of the SC process that plans, implements, and controls the efficient, effective flow and storage of goods, services, and related information from the point-of-origin to the point-of-consumption in order to meet customers' requirements."

SCM is concerned with integrating and managing key processes along the SC. However, managing all resulting process links is likely not to be appropriate; therefore, one can distinguish between

- monitored and managed process links,
- monitored but unmanaged process links, and
- process links that are neither monitored nor managed (cf. Lambert et al. 1998).

There are many other, slightly different definitions of SCM (cf. Cooper et al. 1997; Croom et al. 2000, pp. 68; Kotzab/Otto 2000).

Strategic alliances, networks, and also SC cooperation appear to be attractive options for business governance if the main dimensions in transaction cost theory, namely asset specificity, uncertainty, and frequency show intermediate values. This has been termed the "move to the middle" hypothesis (Clemons et al. 1993).

SCM is aimed at intensive cooperation between companies to improve all intra- and inter-company material, information, and financial flows. Figure 1.1 shows the typical directions of flows; however, in the case of return shipments or refunds different flow directions result.

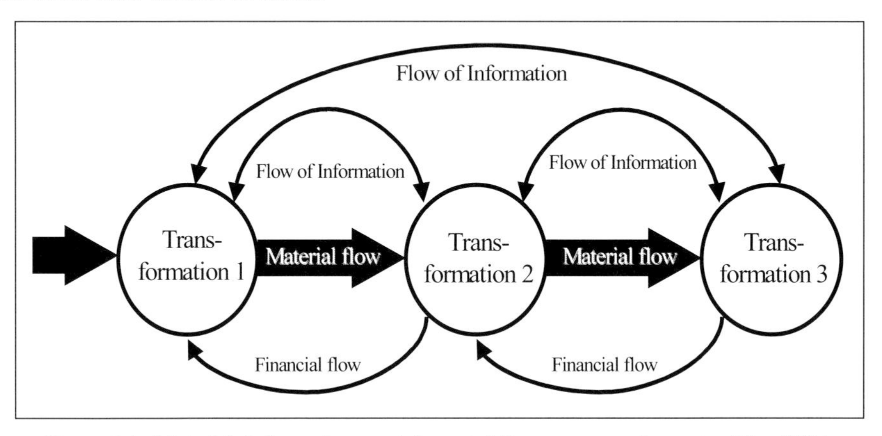

Figure 1.1: Material, information, and financial flows as core elements of the SCM

From a certain position in the SC it is possible to look to the "upstream side" (referring to suppliers) or the "downstream side" (referring to customers). Thus, not only the supply side, but also the demand side are subject to SCM. Depending on the "distance" from the company considered, tier 1, tier 2, up to "tier n" suppliers are distinguished (Committee on Supply Chain Integration 2000, p. 23). Furthermore, the term "chain", on which SCM is based, gives an incorrect view of the realities of business. The idea of a "Supply Network", a "Supply Web", a "Value Net" or a logistics network (cf. Harland 1996; Bovet/Martha 2000) would be more appropriate because a company typically belongs to several SC (cf. Figure 1.2). However, in this book we use the conventional terminology.

Sometimes a distinction is made between primary SC and support SC: Production materials, parts, components, subassemblies, and finished products flow in the primary chains of a manufacturer; inward flows of consumables or MRO (maintenance, repair, operating) items, of capital plant and equipment are viewed as support chains. The latter do not involve outward logistics (Saunders 1997, pp. 151).

A wide range of coordination problems can arise if a company belongs to several SC. However, owing to the close cooperation and the associated duties and organizational measures involved, not all business relationships with other companies can meet the high demands of this organizational model. As when Just-in-Time (JiT) relationships are being built up, the partners of whom such close cooperation will be expected must be chosen with the utmost care.

To reach the goal of customer satisfaction, the SC should function without boundaries; the result is called "Supply Chain Synthesis" and defined as a holistic, continuous improvement process of ensuring customer satisfaction from the raw material provider to the final, finished-product customer (Tompkins 2000, p. 2). However, this idea is not really new; as long ago as 1985, Houlihan formulated it as "... integration, not simply interface, is the key."

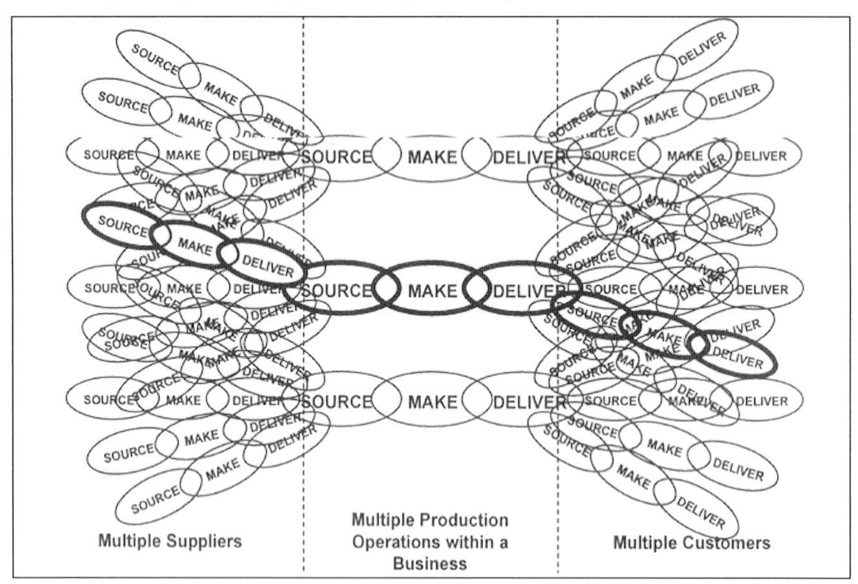

Figure 1.2: Supply Chain as part of a supply network
(based on earlier material from the Supply Chain Council)

Figure 1.3 shows the importance of IS for the support of business processes. Every element in the value chain needs and generates information. Consequently, IS are important for coordination of the divisions and companies cooperating in SC; they are crucial in bridging the traditional boundaries between the members of the chain.

We can assign systems that support SCM (similar to the integrated IS of an manufacturing company) to different levels (Figure 1.3, based on Mertens 2000, pp. 4): The lowest level lists those function areas that we regard as elements of the intra-company value chains when structuring Chapter 2. To meet the tasks arising in these areas, the employees make use of administration and planning systems. Such systems are also known as Enterprise Resource Planning (ERP), operative, or execution systems. SAP R/3 is worldwide market leader in this type of system.

The middle level contains planning, scheduling, and monitoring functions that have been implemented, e.g., by SAP in its Advanced Planner and Optimizer (APO) system. These applications have the following characteristics (based on the components of this system and the mySAP SCM initiative described in more detail in Chapter 3):

- The Supply Chain Cockpit (SCC; cf. section 3.1.1)
- Collaborative Planning, Forecasting, and Replenishment (CPFR; cf. section 3.1.2) and the Collaboration Engine
- Demand Planning (DP; cf. section 3.1.2)
- Supply Network Planning (SNP; cf. section 3.1.3)
- Production Planning and Detailed Scheduling (PP/DS; cf. section 3.1.3)
- The Available-to-Promise (ATP)-function (cf. section 3.1.5)
- The Logistics Execution System (LES ; cf. section 3.3)
- The Alert Monitor (cf. section 3.1.1).

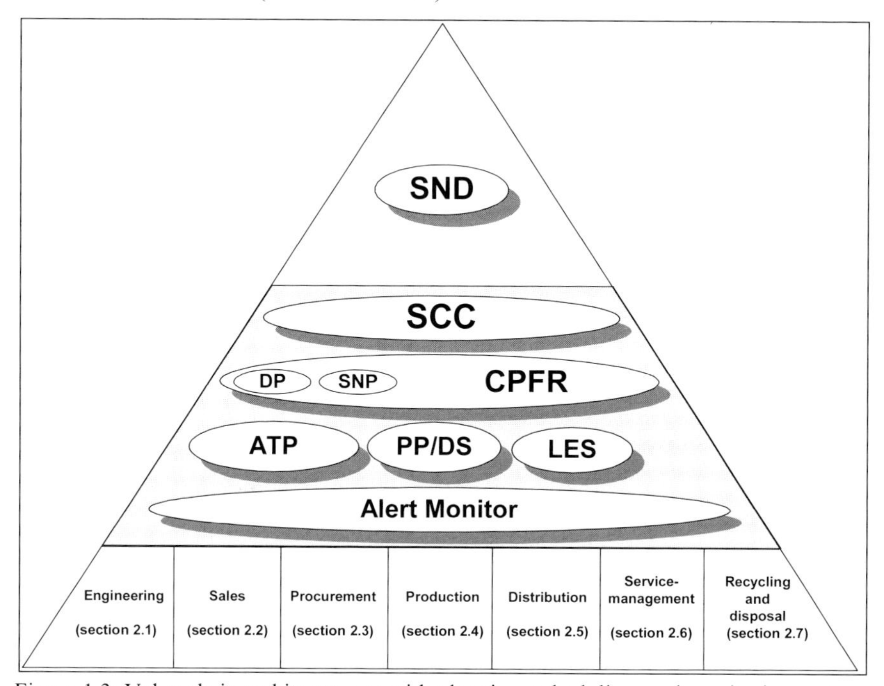

Figure 1.3: Value chain and its support with planning, scheduling, and monitoring systems

The top level of the pyramid contains the Supply Network Design (SND), the strategic planning of partnerships in the SC (cf. section 3.1.1).

SCM affects almost all activities of a company. Table 1.1 provides an overview of the tasks that can be assigned to SCM. From our point of view, SCM concentrates on the form of the cooperation between organizations or organizational units and the associated interfaces. In view of the wide range of these topics, this book focuses on one main part of the subject, namely, IS that may be used to realize SCM concepts.

Table 1.1: Main tasks of Supply Chain Management

Orientation	Strategic	Operative
Internal focus	• Strategies for product and process development • Strategies for providing products and services • Make or buy decisions • Quality management	• Internal quality assurance • Intra-plant transport • Intra-plant storage • Determination of ordering quantities and lot sizes • Optimization of schedules and sequences • Intra-plant IS for planning and controlling of order management
External focus	• Development of an SCM mission • Procurement and marketing strategies • Supplier and customer management • Distribution strategy • Recycling strategy • Definition of an SCM controlling and benchmarking system	• Internet appearance • Research about procurement and sales markets • Evaluation and selection of suppliers • Sales forecasts • Control of the sales force
Dual focus - Pooling of interests	• Supplier and customer structure policies • Coordination of SCM strategies with business partners • Legal basis for SCM partnerships • Joint pursuit of improved business processes	• Managing the organizational and system interfaces • Definition of communication relationships with business partners, paying special attention to IS

1.2 Main Topics in Supply Chain Management

1.2.1 Principles

PriceWaterhouseCoopers (1999, pp. 46) formulated ten principles for SCM. They provide a good overview of the main goals and methods:

1. The prime objective for the SC is to maximize the value to customers and consumers and to the businesses themselves by providing the required level of service at the lowest total cost.
2. Cost and service optimizing should be undertaken across the integrated SC and includes suppliers and customers.
3. Significant cost in the supply chain is associated with non-value-adding activities and the root causes must be understood and eliminated.
4. Excessively sophisticated management solutions to SC problems frequently add still more cost.
5. Demand information and service requirements should be shared upstream with minimum distortion.
6. Synchronizing supply and demand is critical to the service and cost objectives - both in the medium term, to synchronize capacity with market plans, and in the short term, to drive SC activity on the basis of end consumer demand.
7. Reliable and flexible operations are critical to SC synchronization.
8. Integrate with suppliers.
9. SC capacity must be managed strategically; decoupling the supply from demand should be immediately downstream of critical capacity constraints and before significant product differentiation.
10. New product development and new product introduction processes are critical to the performance of the SC.

The requirement listed under point 5, i.e., dissemination of information within the SC, arises out of the "bullwhip effect" (which is also known as "whiplash" or "whipsaw effect"): Relatively small fluctuations in the actual demand among consumers are magnified through the logistics chain and cause larger amplifications, with consequent negative effects on the planning of the production and logistics systems in the earlier stages. "A ripple at one end of the supply chain can trigger a tidal wave at the other" (Fine 1998, p. 91). Figure 1.4 shows the high variance of the resulting order quantities. Sometimes it is assumed that the fluctuations will double in each echelon of the SC (Mason-Jones/Towill 2000).

Four major causes and contributing factors of the bullwhip effect have been identified (Lee et al. 1997a; Lee et al. 1997b):

1. Demand forecast updating
   - No visibility of end demand
   - Multiple forecasts
   - Long lead-time
2. Order batching
   - High order costs
   - Full truckload economics
   - Random or correlated ordering
3. Price fluctuation
   - High-low pricing
   - Delivery and purchase not synchronized
4. Rationing and shortage game
   - Proportional rationing scheme
   - Ignorance of supply conditions
   - Unrestricted orders and free return policy.

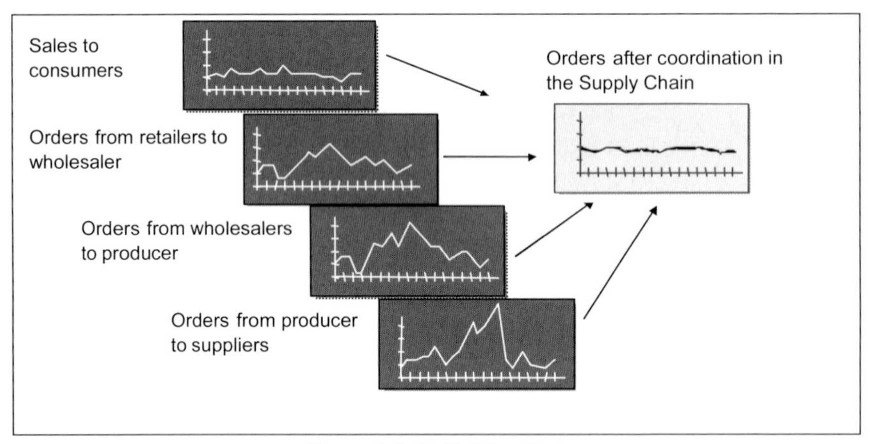

Figure 1.4: Bullwhip effect

Proposals for reducing the bullwhip effect are (Lee et al. 1997a; Lee et al. 1997b; Dornier et al. 1998, p. 223):

- Access to POS data
- Single control of replenishment
- Reduction of lead-times and JiT supply
- Vendor-Managed Inventory (VMI)
- Electronic Data Interchange (EDI) and Computer Assisted Ordering
- Discount on assorted truckload, consolidation by 3rd party logistics
- Regular delivery appointments
- Reduction of frequency and magnitude of special trade deals and sales promotions
- Everyday low prices
- Special purchase contracts
- Allocation of short products based on past sales

- Shared capacity and supply information
- Limitation of flexibility over time, capacity reservation
- Penalties on order cancellations.

Customers can contribute to reducing the order volatility for their suppliers by choosing an appropriate forecasting method. Companies that use smoother forecasting policies impose less of their own volatility on their supply base. Forecasting methods and order policies that respond only to local costs rather than SC-wide costs may be myopic: In the future the company may have to rely on suppliers that have been weakened by earlier, highly volatile ordering policies (Anderson et al. 2000). The widely recommended JiT policies may be harmful from an SCM perspective: In certain conditions the inventory held at a distribution center may have a stabilizing effect and contribute to a reduction of the bullwhip effect (Baganha/Coher 1998); reducing inventory and lead-times in one echelon of the SC may not always improve performance but could result in chaos (Wilding 1998).

The bullwhip effect has become well-known since studies conducted within the Procter & Gamble group (cf. Camm et al. 1997). However, similar "rollercoaster" effects were described much earlier, when the industrial dynamics methods that are based on information feedback control loops were elaborated (Forrester 1958; Forrester 1961). In this context, Forrester (1958, p. 47) mentioned that one way to improve factory production stability is to have reliable retail sales information on which factory production schedules can be based. For this reason, the term "Forrester effect" is sometimes used instead of "bullwhip effect" (Harland 1996), and the Forrester model serves as a benchmark in validating simulation models for SCM (Towill et al. 1992). As early as in 1968, Holt et al. described situations in the TV set industry in which fluctuations in orders tended to increase with progressive movement upstream in the SC, and Houlihan (1985) pointed out the connection between this "amplification effect" and SCM.

In the popular "beer distribution game" the behavior of an SC, consisting of the elements factory, distributor, wholesaler, and retailer, is studied experimentally. Most participants attribute the cause of the frustrating dynamics they experience to external events. Many of them are quite surprised when the actual pattern of customer orders is revealed; few can imagine that their own decisions are the cause of the behavior they have experienced. When asked how they could improve their performance, many call for better forecasts of customer demand that are unknown to the first three elements of the SC (Sterman 1989, pp. 335).

Mathematical analyses indicate that a downstream error is more harmful than an upstream one and that shifting information lead-times from downstream to upstream is beneficial (Chen 1999).

In general, the major cause of the bullwhip effect seems to be a lack of "system thinking" by management. Simulation studies show that the main gains afforded by time compression in the transference of consumer sales data are better order control and reduced stock levels. Reduced distortion and delay of market data through the SC limit the effects of uncertainty resulting from the magnification of orders upstream (Mason-Jones/Towill 1997).

Discovering the bullwhip effect does not automatically solve the problem. The effect can only be avoided if well-established business behavior is altered. Better coordination between companies is typically accomplished by "soft" means such as training as well as by investments in IS. Both of these are expensive, and the expenses must be justified by the benefits in terms of avoiding the unpleasant effect (Metters 1997, p. 99). Simulation models allow experiments with different policies aimed at reducing the bullwhip effect in an SC (cf., e.g., Cachon/Fisher 2000; Carlsson/Fullér 2001).

1.2.2 Tasks and Competencies

The cooperation in the SC generally affects several areas; examples are coordination of procurement and transport strategies, allocation of warehouses along the SC, cooperation with suppliers and customers in the planning process, joint strategies for the outsourcing of functions, and determination of delivery dates with respect to material and capacity availability. Expensive specialized knowledge should be present at only one site within an SC and not in several companies.

The decision between producing in-house and/or buying from an external supplier of products ("make or buy") is often made after consideration of just the short-term effects on costs or cash flows. From this perspective, purchasing from an external supplier appears to be a means of overcoming short-term bottlenecks by so-called premium capacity. For this reason, in addition to the traditional cost and financial accounting methods, a bottleneck-oriented evaluation and thus a charging of opportunity costs also become relevant. Various decision methods are available to support outsourcing decisions (Mertens/Knolmayer 1998, pp. 21). Many (particularly medium-sized) companies retain third parties to operate their ERP applications. In future, the use of third-party-operated ERP and SCM software over the Internet will increase in importance ("Application Service Providing"; ASP; cf. SCN Education 2000).

One form the cooperation in a logistics chain can take is for two or more companies to make some of their information available to each other. Mutual access to existing IS or implementation of new systems with a common structure for joint use may be established. This requires joint activities of the companies cooperating in the SC. These activities may be organized as legally independent organizations (e.g., by transferring some or all IS of the SC to them). Specialized outsourcing service providers may also undertake these tasks.

Among other things, Business Information Warehouses (BW) are used in large enterprises to link IS that evolved in time and are difficult to integrate. SC warehouses can be used even for heterogeneous system architectures in the individual elements of the SC for quick realization of a comprehensive information exchange. In such warehouses, the operative data of the individual partners are saved as extracts and/or in compressed form. The datasets present in ERP applications should be processed and transferred appropriately for this purpose. The agreement of the companies cooperating in the SC must cover a common data model, a joint terminology (semiotics), the transformation rules from the operative systems, and the deletion of obsolete data.

Whereas the use of third parties for operative services (e.g., in the areas of advertising, legal and tax advice, or recruitment) has long been practiced, a tendency has appeared in recent years to outsource tasks that had traditionally been considered to be strategic and were thus thought to be less suitable for outsourcing. These include, for example, development work, which, because of the advantages that medium-sized companies may have in innovation capability compared with large enterprises, is increasingly outsourced to external specialists. Another option is a cooperation with SC partners during product development (cf. section 2.1.1). SC brokers can coordinate services along (sub)chains. Recently there have been discussions on outsourcing not only of individual tasks, but also of comprehensive subsystems of the value chain to third parties.

SCM changes the competitive situation from competition between individual companies to one between different chains of companies. Widely accepted recommendations of strategy consultants suggest concentrating on core competencies. All other tasks can be better performed by third parties, and these are at least candidates for partial or complete outsourcing. SCM raises the question of the organizational unit to which the core competencies should be assigned:

- Core competencies within the considered enterprise
- Core competencies of the SC (outsourcing from the enterprise considered to partner enterprises in the SC)
- External competency (outsourcing to enterprises that are not integrated in the SC).

It must be decided by the management of each company whether core competencies that partially overlap within the SC should be abolished. This policy can cause problems, particularly if the SC becomes unstable.

1.2.3 Interfaces and Complexity

Division of labor is a phenomenon central to developed societies and economic systems. However, the acceptance of this concept from the macroeconomic perspective should not distract us from the knowledge that from a business viewpoint there are different ideas about the optimal degree of the division of labor. On the one hand, division of labor means that the individual tasks can be performed more economically. On the other hand, organizational and system interfaces (cf. Figure 1.5) that interrupt the material, information, and financial flows cause coordination costs between various functions, companies within a group, and legally independent enterprises. Interfaces are a potential source of inefficiency and waste of resources; they require agreements, interrupt information flows, thwart overall planning, cause delays and lack of transparency, and result in transaction costs (Horváth 1991).

Processes within the individual function areas no longer represent the decisive rationalization potential, but rather the interfaces between the business functions and especially between business partners. Recent developments in information and communication technology provide a potential for bridging these gaps more economically.

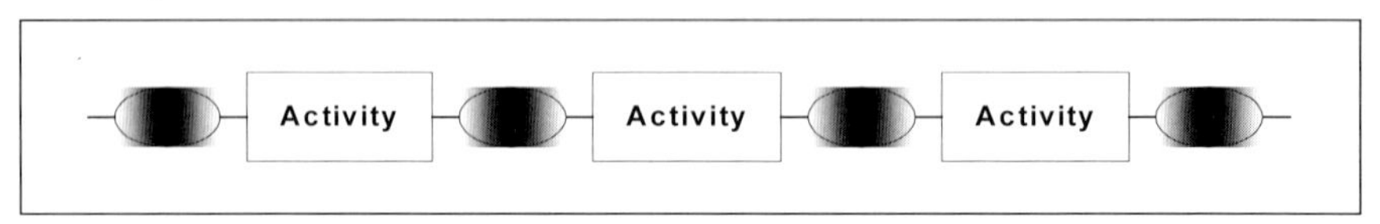

Figure 1.5: Division of labor and interfaces in business processes

In contrast to the recommendations for reduction of specialization and thus of the number of intra-company interfaces, a division of labor over company boundaries implies basing services on the group's own core competences. Consideration of the total number of work steps needed to make a product reveals a tendency to reduce the number of interfaces within the organization and, at the same time, to increase the number of external interfaces. This may result in fewer interfaces, but these are typically more difficult to master (cf. Figure 1.6).

One possible explanation for these divergent approaches to interfaces is that IS have long been used within companies to improve information exchange between different functions. In contrast, the data exchange between companies is usually restricted to the transfer of routine data; management information is seldom exchanged. Consequently, a simultaneous reduction in the number of interfaces that exist within companies and an increase in the number of interfaces to be handled between companies can be desirable if more intensive use of the information technology (IT) potentials for cross-company cooperation is aspired to. Precisely this is one of the important goals of SCM.

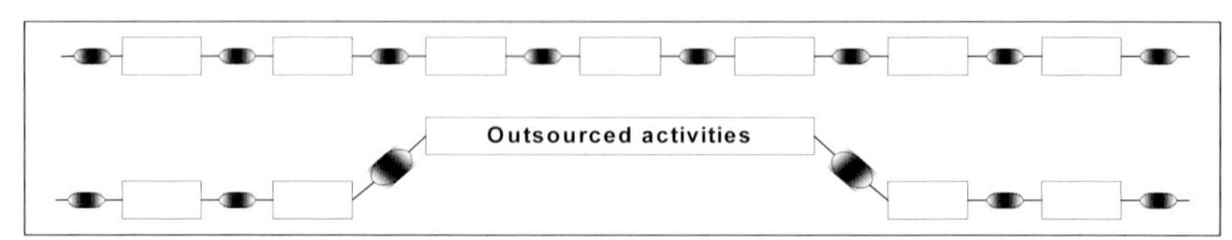

Figure 1.6: Internal and external interfaces for partial outsourcing of tasks

In many enterprises, user department employees are responsible for specific applications. These often exchange information with several other applications or systems. The persons responsible for such interfaces are usually not named explicitly. For instance, if two systems that exchange data with each other are maintained in parallel, care must be taken to ensure that the interfaces remain functional. Furthermore, some middleware concepts assign not only technical but also business functions to the interfaces. Consequently, the areas of responsibility for their development and maintenance should be clearly defined and documented.

EDI, ERP systems, and SCM software are forcing companies to adopt common practices, e.g., for handling of bills of materials (BOM), shipping, invoicing, and accounting. As logistic transactions become more and more standardized, exchange of information and measurement of performance across the SC will become easier (cf. Keebler et al. 1999, p. 223).

In the SAP environment, the wide range of bilateral interfaces will presumably be narrowed by using the Internet in conjunction with the "Business Framework" concept (Figure 1.7). With the Business Framework Architecture SAP defines a conceptual and technical frame in which independent, configurable software components can be embedded and within which they can communicate with each other. Major components include standardized programming interfaces, the Business Application Programming Interfaces (BAPIs; cf. Moser 1999). This interface technology permits simple access to the business functions of the R/3 System. Examples for BAPIs are the creation or change of a customer order or the availability check for specific products at the warehouse of a certain supplier. The components also enable functions of the R/3 System to be made available over the Internet (cf. Buxmann/König 2000, pp. 84).

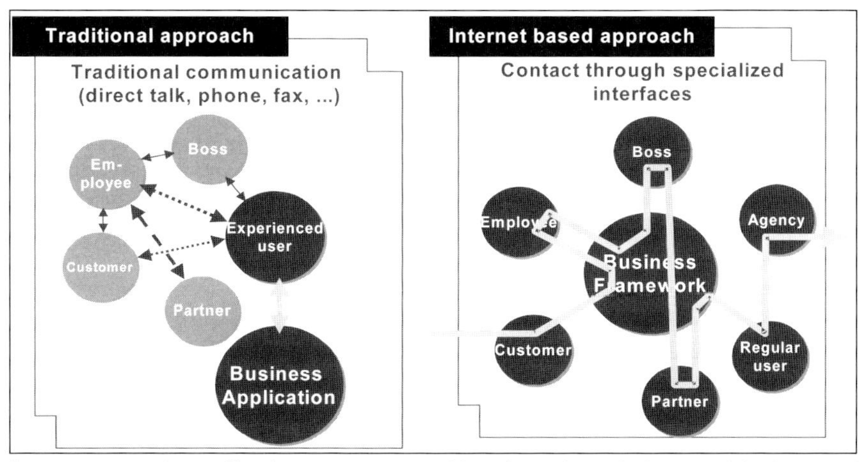

Figure 1.7: New form of business processes and interfaces through the use of the SAP Business Framework and of the Internet (cf. Bussiek/Stotz 1999)

Globalization may add complexities and uncertainties to SC by

- explosive dimensions of product variety,
- substantial geographic distances,
- additional forecasting difficulties and inaccuracies,
- exchange rates and other macroeconomic uncertainties, and

- infrastructural inadequacies
  - insufficient worker skill
  - missing supplier quality
  - lack of local process equipment and technologies
  - inadequacies in transportation and telecommunications infrastructure (cf. Dornier et al. 1998, pp. 224).

One of the main goals of SCM is improved mastery of the complexity of intra- and inter-company cooperation (cf. Barratt et al. 1998; New 1998). This can be a-chieved by implementing comprehensive planning and control systems and/or by reducing the complexity of the real system (cf. Figure 1.8). Sometimes it is argued that the evolution of IS has created additional complexity in real systems (Wilding 1998; Fernández-Rañada 1999, pp. 167). Reduction of the complexity in the cooperation both within and beyond company boundaries is an important goal of SCM. It is not rationalization at the expense of other SCM companies that is aspired to, but rather win-win situations for the cooperating enterprises.

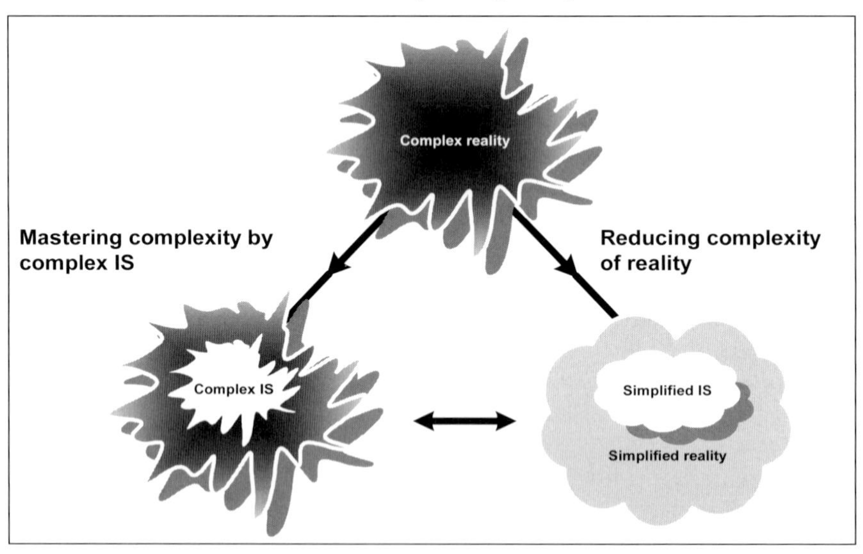

Figure 1.8: Possible ways of mastering complexity

Reduction of the complexity of real systems, as emphasized by Burbidge (1962) since the early 1960s, is a major idea in reengineering an SC. Thus, main concepts of SCM were stimulated by the work done on system dynamics by Jay Forrester and on complexity reduction by Jack Burbidge; this led to the artificial catchword FORRIDGE (cf. Towill 1994; Towill 1997). Table 1.2 summarizes the means of complexity management that are relevant to reorganizing SC processes.

Table 1.2: Starting points for reducing complexity by SCM (Hughes et al. 1998, p. 111)

Harmonize: Standardize: Rationalize	Harness the Electronic Supply Chain	Challenge Low Added Value Work
• Brand rationalization • Stock keeping unit reduction • Integrated specifications and systems platforms • Process design • Close supplier involvement • Supplier reduction and rationalization • Integrated design from R&D onwards	• On-line trading • Computer connectivity • Electronic Data Interchange (EDI) • Transaction processing centers • On-line catalogs • Self-directed purchasing • Electronic funds transfer • Internets and intranets • Efficient Consumer Response (ECR)	• Outsourcing non-core work • Use preferred suppliers • Supplier planning • Certification programs • Supplier self-monitoring • Develop material scheduling • Vendor Managed Inventory (VMI) • Adopt purchasing cards

Twelve "Requirements of Success" have been formulated for achieving simplified manufacturing processes (Tompkins 2000, p. 236):

1. Simplified product design
2. Reduced lead-time
3. Reduced production lot sizes
4. Reduced uncertainty
5. Balanced, focused departments and factories
6. Straightforward and transparent production and inventory control systems
7. Reduced inventories
8. Increased adaptability
9. Increased quality
10. Reduced down-times
11. Continuous-flow manufacturing
12. Upgraded tracking and control systems.

It is suggested that the SC be simplified wherever possible. Such a redesign may be supported by modeling existing and proposed SC architectures and by using dynamic simulation to evaluate alternative options (cf. Towill et al. 1992). In this context, the following 12 rules or recommendations for simplifying material flow have been formulated (cf. Towill 1999):

1. Only make products which can be quickly dispatched and invoiced to customers
2. Only make in one period those components that are needed for assembly in the next period
3. Minimize the material throughput time, i.e., compress all lead-times
4. Use the shortest planning period, i.e., the smallest run quantity which can be managed efficiently
5. Only take deliveries from suppliers in small batches as and when needed for processing or assembly
6. Synchronize "time buckets" throughout the SC
7. Form clusters of products and design processes appropriate to each value stream
8. Eliminate process uncertainties wherever possible
9. Understand, document, simplify, and only then optimize the SC
10. Streamline and make highly visible all information flows
11. Use only proven simple but robust decision support systems
12. The business process target is the seamless SC, i.e., all elements of the chain have to think and act as one organization.

1.2.4 Cooperation

In addition to the topics handled in logistics, SCM is concerned with the conditions that permit enterprises to achieve a long-term, stable, and economically successful cooperation. Significant for establishing a successful SC is a trusting cooperation between the managers and the staff of the various organizations. Trust is primarily person related and thus dependent on competence, experience, and personal behavior of the staff members. In addition to personal relationships, it is important for the functioning of SC that the organizations also build up trust in each other and team up in their daily activities. For example, it has been suggested that SCM partners should publish a joint newsletter (Underhill 1996, p. 259) and thus make their cooperation visible to the markets and to the public. Another recommendation is that customer and supplier round tables be held to provide an opportunity of sharing ideas interactively, querying existing beliefs, and helping new opinions to emerge (Tompkins 2000, p. 212).

Lack of trust is considered to be a major reason why the implementation of SCM concepts in practice is still lagging significantly behind the proposals developed in theory and recommended by consultants (Monczka et al. 1998, p. 85). The following factors should be considered when a relationship of trust is to be built up (Siemieniuch/Sinclair 2000, pp. 263):

- Identification of common goals and policies
- Transparency about problems and ways of working
- Willingness to share benefits
- Common understanding of terms and their usage

- Respect for confidentiality
- Efficient execution of promises
- Personal relationships, built over time
- Recognition of a "favor bank" with credits assigned to persons who have bent the rules to resolve a severe problem.

From another perspective, the basic ingredients for successful logistical alliances are described as follows (Daugherty 1994, p. 763):

- Thorough assessment of principal strengths
- Consistency of values
- Clear understanding of objectives
- Agreement on measurement standards
- Long-term focus
- Commitment at multiple levels
- Working relationship at interface level
- Limited number of relationships
- Elimination of "quid pro quo"-mentality
- Negotiated prices - not bid and subject to change.

SC relationships should have a long-term perspective. If the members of an SC work together only for a short time, the investments made in building it up cannot provide an adequate return. Should individual members leave the SC, not only they, but also the remaining members of the SC can incur high switching costs.

Problems may result from providing misinformation within the SC (cf. Cachon/Lariviere 2001) to obtain an individual advantage or from the misuse of information obtained confidentially. The "business partners" may use information they receive to achieve competitive advantages over the companies that have provided data in good faith. For example, knowledge about high inventory levels of a supplier may result in demands for price reductions. It is usual to agree that only a selected group of employees of an SC partner has access to such data and has to treat them confidentially. There is also the problem of ensuring that third parties cannot access sensitive data.

In business processes that have evolved over time, information flow is often fragmented because it is the result of successive implementations of applications designed for stand-alone use (cf. Jones/Riley 1985). In reorganization projects this information flow is often designed with fewer interruptions. In addition, process analyses can make it possible to avoid certain functions or to reassign the work steps to different elements in the SC. A typical example involves the quality control traditionally performed by both the supplying and the purchasing company. A close cooperation in the SC can allow the purchasing company to cease making these tests in certain circumstances. Different forms of payment are also possible in which the supplier can initiate the transfer of financial assets; an example of this is the "Cash in Time" system implemented by ABB. Such cases emphasize that the realization of SCM concepts may change the internal business processes (cf.

Blackwell 1997, pp. 125), lead to the definition of new business rules, and reduce bank transaction charges, but may also lead to new organizational units responsible for solving disputes between members of the group or the SC.

Whereas EDI-transferred data satisfy the information requirements of the partners only partially, access to some parts of a partner's IS can provide relevant information early, e.g., for forecasts, inventory leveling, use of resources, order tracking or shipping status (cf. Figure 1.9).

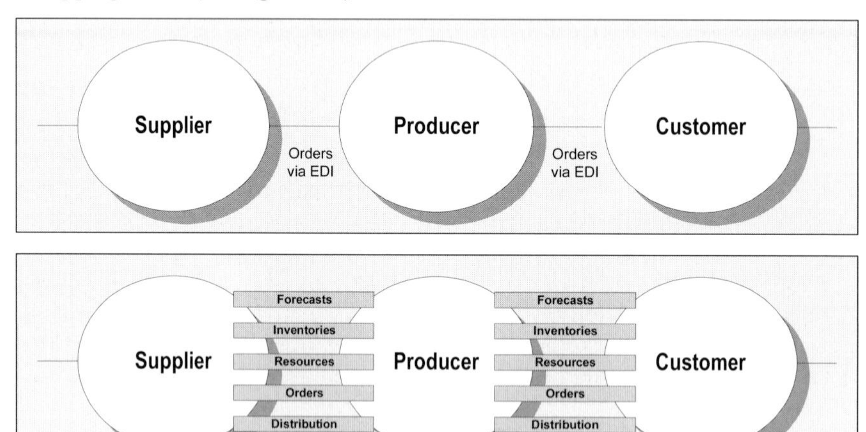

Figure 1.9: Extension of the information exchange in SCM

A questionnaire responded to by large American enterprises indicates that significantly more data are made available to the business partners now than a few years ago (Table 1.3).

Table 1.3: Percentage of large US companies that make specific types of data available to their business partners (PriceWaterhouseCoopers 1999, p. 34)

Type of data made available	**1998**	**Forecast 2001**
Inventory and capacity	50%	75%
Demand history and forecasts	30%	72%
Order status	30%	66%
Project design and specifications	34%	54%
Financial information	3%	20%

In addition to information exchange, cooperation within the SC must lead to coordinated and concerted decisions. The SC aims at providing a global optimum solution; consequently, SCM is a concept that requires (logical) centralization (cf. Christopher 1998, p. 258). Because the achievement of the global optimum does not necessarily lead to an individual optimum for each member of the SC, a company following the SC goals can find itself in a worse situation than one that would result from realizing its optimal solution outside the SC. For example, an agreement in the SC requires that a company must accept a suboptimal supply rhythm and thus increased inventory levels, because it is subject to delivery and route planning that gives preference to the interests of other companies. In some situations the SC will have to compensate the disadvantages to cooperation partners that have to accept discrimination of this kind. This results in an accounting problem similar to that encountered in settling the costs of coupled products.

In an SC, the companies may use different Material Management Systems (Base Stock Control, MRP, Line Requirements Planning, Distribution Requirements Planning). One goal of the SC is to create networked inventory management systems (NIMIS). A group of NIMIS should perform like one integrated system for inventory management across the networked companies (Verwijmeren et al. 1996).

Because of the simpler form, cooperation in an SC often begins between different plants and subsidiaries within a group. The provision of information previously available only in decentralized units results in an advantage to the partners in the chain and thus for the whole group. It must also be decided within the group how to compensate individual areas of responsibility for the effects that result from taking an overall perspective. Thus, the implementation of a group-internal SCM can change the level of the settlement prices or even the system for determining them.

1.2.5 Types of Supply Chain Management

SCM concepts can be relevant at various organizational levels:

1. SCM between different plants of an enterprise or between different companies within a group

The organization of logistics in decentralized enterprises deserves special attention because, from the group viewpoint, much information (e.g., on eliminating used assets or on inventory levels) is often only locally available and this often results in unsatisfactory decisions. In particular, the use of stocks available at affiliated companies or subsidiaries is impossible when inventory information cannot be exchanged in real-time. Consequently, some companies have started to centralize such information at least in the logical layer of data modeling. Without such a centralization strategy complicated issues of designing incentives schemes in the SC arise (Lee/Whang 1999).

Figure 1.10 shows how the intra-company workflow is embedded in a comprehensive inter-company SC.

2. SCM between two companies in neighboring positions of the value chain

The coordination in an SC can be based on bilateral agreements between neighboring companies in the chain. These organizations have internal IS to support their internal processes. Often these systems also communicate beyond enterprise boundaries to a limited extent. This means that there are many bilateral agreements, industry standards, and also inter-industry standards, e.g., for the exchange of order-related data via EDI or of design data. The aim of an automated data interchange is to avoid entering data more than once and to process digitally available data further in other units or organizations. The problems that arise with the exchange of sophisticated data, in particular between heterogeneous IS, should not be underestimated.

Nowadays, the realization of the concept of inter-company data exchange is mainly aimed at providing an IS interface between *two* organizations. The traditional concepts ignore the broader cooperation along the value chain.

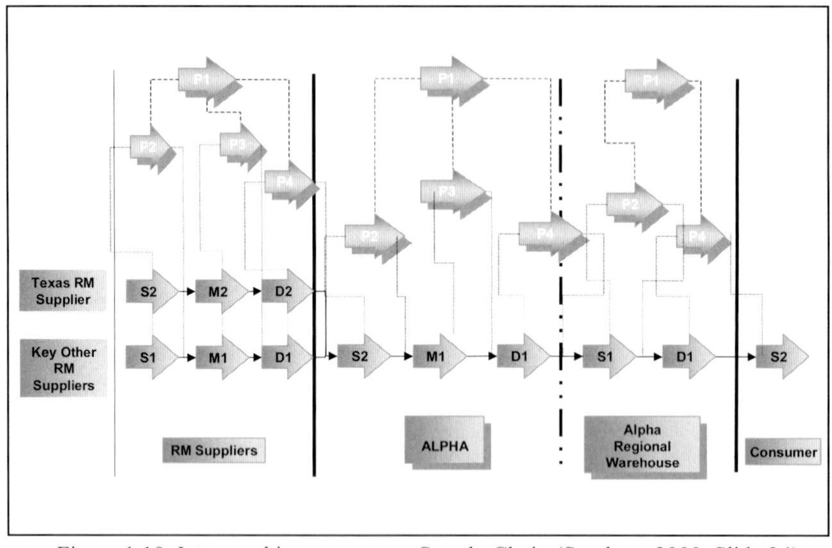

Figure 1.10: Intra- and inter-company Supply Chain (Stephens 2000, Slide 24)
Legend: D1 .. Other suppliers; D2 .. Deliver Make-to-order Products; M1 .. Make-to-stock; M2 .. Make-to-order; P1 .. Plan Supply Chain; P2 .. Plan Make; P3 .. Plan Source; P4 .. Plan Deliver; RM .. Raw Material; S1 .. Source Stocked Material; S2 .. Source Make-to Order Products

3. SCM between more than two companies

In the long run, a cooperation and, thus, a coordination over several echelons of a value chain is more promising than bilateral agreements. In this case, usually one company, the "shaper" of the SC, takes the initiative to motivate the upstream and downstream business partners to implement a complete SCM concept.

The data relevant for the SCM can be collected in a BW and made available (differentiated according to access rights) to the companies involved in the SC. Such a (at least logically) centralized "Supply Chain Information Warehouse" largely decouples information and material flows. The SC Server becomes a central point of the information supply for the companies involved in the SC. The data supplied by the partners can also be augmented with non-SC contents, e.g., from business services, information brokers, online databases or the Internet.

1.2.6 Methodology of the Supply Chain Council

The Supply Chain Council [http://www.supply-chain.org/] has been in existence since June 1997 and had about 800 members in spring 2001. The council illustrates the SC with Figure 1.11, in which, according to the Supply Chain Operations Reference (SCOR) Model, planning, sourcing, production, and delivery appear as main processes. In October 2000, version 4.0 of the SCOR model was released. The most significant change in this release was the inclusion of "Return" as a main process; this expansion extends the scope of the model into the area of post-delivery customer support.

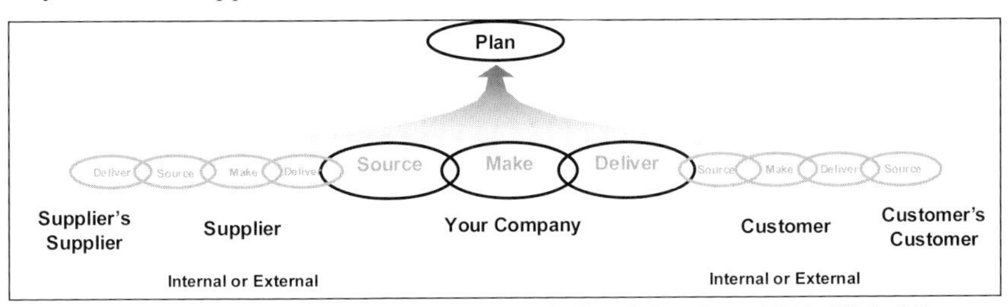

Figure 1.11: Visualization of the supply chain (cf. Stephens 2000, Slide 12)

Figure 1.12 shows the procedure model recommended by the Supply Chain Council, also differentiating between intra- and inter-company cooperation.

1.2.7 Potential Benefits and Controlling

The SC Council sees SCM providing the following major potentials for improvement (Stephens 2000, Slide 39):

- Increase forecast accuracy by 25 to 80 percent
- Reduce inventory levels by 25 to 60 percent
- Reduce fulfillment cycle time by 30 to 50 percent
- Lower SC costs by 25 to 50 percent
- Upgrade fill rates by 20 to 30 percent
- Improve delivery performance by 16 to 28 percent

- Meliorate capacity realization by 10 to 20 percent
- Improve overall productivity by 10 to 16 percent.

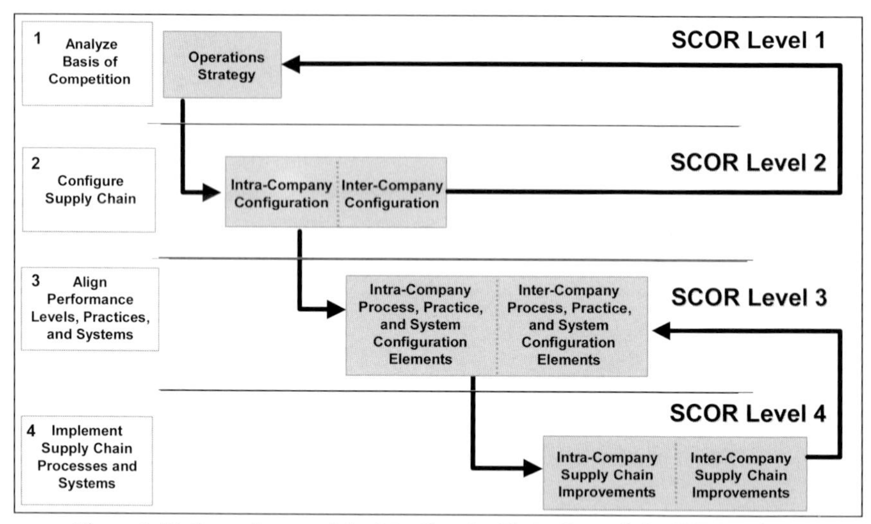

Figure 1.12: Procedure model of the Supply Chain Council for SCM projects (cf. Stephens 2000, Slide 20)

According to the results of a survey published by Deloitte Consulting in spring 1999, more than 90% of North American companies consider SCM to be crucial or particularly important for their success. Whereas only two percent of the polled companies designate their SC as "world class", 73 percent evaluate it as average or below average. Formal SC initiatives had been completed, were in progress or planned in 90 percent of the companies (Deloitte Consulting 1999b).

Evaluation of the economics and benchmarking require SCM-specific key figures. Examples are quality defects, inventory, new product development cycle, order cycle time, investment in tooling, manufacturing space, and distribution costs (cf. Hughes et al. 1988, pp. 97, pp. 201). Controlling logistics should be extended to SC controlling (Zäpfel/Piekarz 1996). The "Key performance indicators" can be represented in the form of "Supply Chain Scorecards" (cf. Christopher 1998, pp. 101; Ptak/Schragenheim 2000, p. 128). A checklist of 72 questions, structured in 9 audit points, for assessing an SC has been compiled by Tompkins (2000, pp. 240) and guideline checklists for the supply management core processes are proposed by Riggs/Robbins (1998, pp. 151).

The Supply Chain Council defines the main metrics in the SCOR model as

- Delivery performance
- Fill rate by line item
- Order fulfillment lead-time
- Perfect order fulfillment
- Supply chain response time

- Upside production flexibility
- Supply chain management cost
- Warranty cost as percentage of revenue
- Value-added per employee
- Inventory days of supply
- Cash-to-cash cycle time
- Asset turns (Ptak/Schragenheim 2000, pp. 126).

In many cases measures are defined jointly by suppliers and customers. A detailed description of the present status in measuring the effects of SCM is presented by Keebler et al. (1999), who also propose the following rules of effective SCM measurement:

1. Start with strategy
2. Understand customer needs
3. Understand other drivers that influence measures
4. Adopt a "process" view of logistics
5. Focus only on key processes and activities
6. Designate a "process owner"
7. Use only a few key measures
8. Steer clear of subjective measures
9. Avoid excessive averaging
10. Understand the statistics (Keebler et al. 1999, pp. 99).

Experiences gained at Hewlett-Packard (HP) show that a clear understanding of business processes and the definition of robust metrics are essential to improve order fulfillment in an SC. The distribution function was carefully measuring its internal ability to process customer orders, but never compared shipment performance to customer delivery requests. HP realized that the only way to take control of the order fulfillment process was to revisit the fundamentals of measurement and control. A measurement scheme was proposed which tracks simultaneously the relationship between service and inventory while monitoring and managing delivery reliability (Johnson/Davis 1998).

1.2.8 Possible Obstacles

On the journey to mature, "ideal" SCM systems a number of obstacles and dangerous areas must be overcome:

1. The many agreements needed in an SC could make the cost of changing partners prohibitive. The SC would then become a relatively fixed structure that cannot exploit market chances in a sufficiently flexible way, e.g., by including new, aggressive suppliers with an innovative product spectrum. The associated strengthening of the barriers to entering the market can be undesirable from a macroeconomic point of view for competitive and innovation reasons, and the institutionalization of SC may be impeded by anti-trust legislation.

2. Small and medium-sized companies with inadequate personnel and/or financial resources could be forced into SC systems by larger business partners even if their resources are insufficient for this type of cooperation.
3. As discussed in section 1.2.4, the optimum for the supply network does not necessarily match with the individual optimum for each participant: Some partners achieve more and others less than proportional advantages from the SC and some may be even worse off. The quantification and the apportionment of the effects and the specification of compensation mechanisms provide considerable potential for conflict.
4. Scheduling of the material flows influences the achievement of several objectives, e.g., low inventory levels versus short delivery times, cost reduction by a smaller number of warehouses versus short transport distances. The question arises as to whether this complexity can be adequately represented in computer-supported scheduling systems, e.g., by rule- or knowledge-based components, and whether the maintenance effort associated with them is acceptable.
5. It has been clarified only to a limited extent whether it will be possible to make software products for SCM suitably chain neutral. It is possible that some properties of the industries are so different that the packaged software based on the traditional integration paradigm cannot handle them. The extent to which new software architectures, such as componentware, may overcome these problems cannot presently be foreseen.
6. The implementation of a conventional ERP or warehouse management system is an intra-company project for which "only" units that belong to the same management hierarchy need be coordinated. In contrast, SCM requires that project portfolios, resources, priorities, and plans are coordinated between several independent companies. Unwillingness or delays in one company can affect the activities of the partners and, in the worst case, cause reactions within the SC.
7. The development and implementation of SCM software is a major undertaking, with all the associated risks relating to the observance of design goals, deadlines, and costs. Sometimes an analogy to the concept of Computer Integrated Manufacturing (CIM; cf. Scheer 1994) is made, with the implied fear that SCM could fail in a similar way, because of excessive expectations and the complexity of its realization.
8. One might argue that the trend to E-Business and B2B marketplaces makes switching between several business partners easier and that this trend could lead to an "end of the supply chain" (Singh 1999).

Figures 1.13 and 1.14 show results of a global empirical study on the main difficulties encountered with SCM.

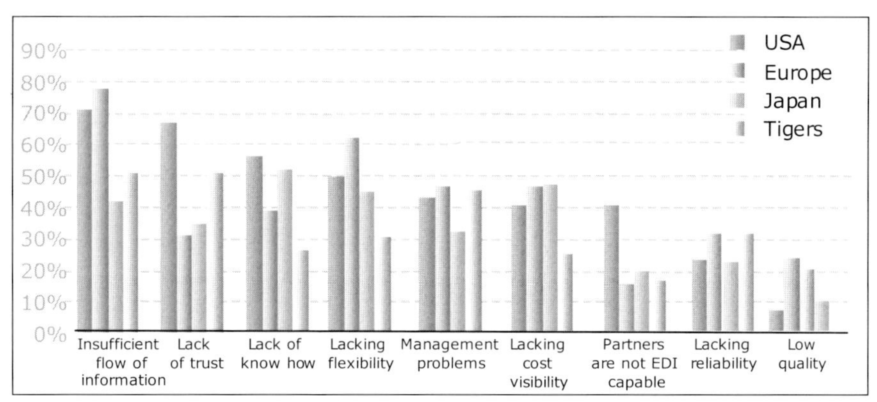

Figure 1.13: Main difficulties concerning the integration of supply chain partners (Baumgarten/Wolff 1999, p. 61)

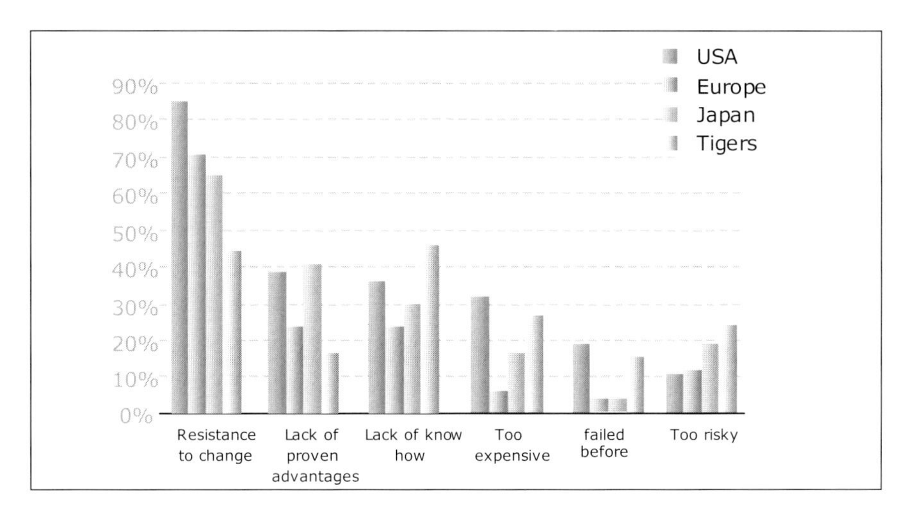

Figure 1.14: Barriers when attempting to reengineer supply chain processes (Baumgarten/Wolff 1999, p. 61).

Consequently, in addition to many euphoric forecasts of remarkable growth of the SCM market, attention should also be paid to the skeptical voices of those who see the SCM as a "bubble" waiting to burst. Other sources discuss the "traps" and challenges that accompany the journey to a successful SCM (Lee/Billington 1992; Committee on Supply Chain Integration et al. 2000, pp. 66). With reference to this situation, the reader is referred to Chapter 5, which provides an overview of the methods applied and the experience recorded by pilot users of SCM systems.

1.3 Software Support for Supply Chain Management

1.3.1 Enterprise Resource Planning Systems

ERP systems resulted in dramatic changes of many intra-company information architectures in the 1990s. SAP AG, with its R/3 product, is the unchallenged leader in this market segment and has become one of the largest and most influential software vendors in the world.

According to an opinion poll conducted by InformationWeek, more than 50 percent of the IS managers are involved in building a "business supply chain" that provides business partners with online access to their ERP system (Stein 1998).

In the previously mentioned poll by Deloitte Consulting, 80 percent of the companies judged IS support for the SC as critical or very important. Many companies expressed disappointment with the limited functionality offered by ERP systems for the management of SC and doubted whether ERP systems should be regarded as creators or destroyers of competitive advantages (cf. PriceWaterhouseCoopers 1999, pp. 63). Without an overarching strategy for the IS facilitating the SC, the benefits of such investments are bound to be isolated and will not have a profound effect on the SC overall. Only 17 percent of the polled companies were satisfied with the installation of just an ERP system whereas 41 percent plan to use it only as a transaction backbone and make various extensions that also affect SCM (Deloitte Consulting 1999a; Deloitte Consulting 1999b). In particular, the inadequate decision support provided by ERP systems is criticized.

Integration and interfaces are important issues in SCM and thus also of software supporting SCM. In 1995, several ERP vendors and other companies founded the Open Applications Group [http://www.openapplications.org/]. In 2001, the group had about 85 members, including SAP, Oracle, and PeopleSoft. Its mission is to define and encourage the adoption of a unifying standard for E-Business and application software interoperability that reduces customer cost and time to deploy solutions. Development of a common model should mean that the end-users would have an easier, less risky transition from existing applications to new applications with near "plug and play" integration. Instead of traditional point-to-point connections, the Group recommends a common integration backbone strategy. In 2001, Release 7.1 of the OAGIS (Open Applications Group Integration Specifications) was published, which is aimed at accelerating component integration and E-Commerce (EC) by adding capabilities for SC Integration and the use of the eXtensible Markup Language (XML). The integration scenarios include frameworks between

- Manufacturing and Financials
- Manufacturing and Order Management
- Manufacturing and Billing
- Manufacturing and Shipping
- Manufacturing and Purchasing
- Manufacturing and Human Resources
- Sales-Force Automation and Order Management
- Sales-Force Automation and Shipping
- Sales-Force Automation and Purchasing
- Customer Service and ERP
- Customer Service and Financials
- Financials and Human Resources.

1.3.2 Supply Chain Management Systems

Sometimes a distinction is made between IS that support Supply Chain Planning (SCP) and those supporting Supply Chain Execution (SCE). Systems for advanced intra- and inter-company planning are assigned to the first product group, and systems for data management and communications to the second. SCP includes the decision-making process and the analysis tools, forecasting algorithms, data filtering tools, and other decision support techniques. These combine to produce management information about what and when materials and services are needed from a company's suppliers and when to satisfy the demand from the company's customers. The vendors of ERP systems are entering the SCP market by adding Advanced Planning and Scheduling (APS; cf. sections 2.4.2 and 3.1.3) features to their existing offerings. Market leaders in SCP systems are (in alphabetical order) i2 Technologies, Manugistics, Oracle, and SAP.

SCE includes the process and technology for communicating and executing the decisions that result from SCP. The SCE software market is highly fragmented because of the industry-specific business needs that must be addressed for a satisfactory SCE solution. Some vendors of SCE tools are the Descartes Systems Group, EXE Technologies, Industri-Matematik International, Manhattan Associates, and Provia Software (Bjorksten et al. 1999).

Philippson et al. (1999) provide an overview of 15 SCM products, and Szuprowicz (2000, pp. 117) also summarizes the properties of several SCM software packages.

The software solutions offered by these companies must allow connections to several ERP systems from which specific aggregate data of transactions currently in progress or executed earlier can be obtained. If these connections exist only as file transfers, the optimization algorithms and rule bases of SCM heuristics are only loosely connected with the operative systems.

In mid-1997 SAP announced various SCM systems, which were initially grouped under the SCOPE (Supply Chain Optimization, Planning, and Execution) acronym. However, SAP later used the term "SCM Initiative" as one of several "New Dimension Initiatives". Meanwhile SAP presents SCM as one of its mySAP solutions.

Other vendors of ERP systems also provide SCM functionality and use a similar nomenclature. For example, Baan developed a concept called Baan SCS (Supply Chain Solutions); J. D. Edwards calls its SCM system SCOREX (Supply Chain Optimization and Real-Time Extended Execution). In Oracle Applications, one of many modules is designated "Supply Chain Management". XRP (EXtended Resource Planning) is sometimes also used as an acronym for SCM software (Coomber 2000).

One obvious procedure would be the attempt to identify and implement the most suitable software solution for each relevant SCM functionality. In this case users build up their SCM concept using "best-of-breed" products that best satisfy globally formulated requirements from the static perspective at a given moment. As a manager at National Instruments argued in 1997 with respect to the dynamics of system properties: "The best-of-breed software packages have more functionality today, but that might not be the case tomorrow" (Stein 1997). These architectures require major integration and maintenance work by the user. Such work may repeat changes to the system components with each release. The associated problems lead many users to prefer one-stop shopping even if each individual function is not supported in the best possible way. The concept to build up applications using several modules that may be delivered from different sources (componentware principle, "software composition") and to standardize the interfaces better than previously may lead to a reorientation of users in the long run. More capable system integrators ("processware companies") that provide special "off-the-shelf" integration services as part of ERP systems could result in higher acceptance of the componentware concept. The integration problems become evident when even competent consulting firms and system integrators specialize in the implementation and integration of a few systems because high learning efforts are necessary if other products have to be integrated.

SAP follows the componentware concept with its Business Framework. The integrated components do not necessarily need to stem from R/3 or the same R/3 release. Business Components and Client Components can be integrated using Application Link Enabling (ALE) or via SAP Business Workflow.

Currently, it appears that SCM products of ERP vendors that are supported with regard to integration and maintenance are receiving more interest than isolated SCM packages. In addition, the future perspectives of the niche providers of SCM software are sometimes judged critically. For example, J. D. Edwards, an ERP vendor, bought Numetrix in 1999; three years ago, PeopleSoft acquired Red Pepper, another vendor of SCM software.

Although the ERP suppliers claim that open interfaces would provide a good potential for the development of, e.g., industry-specific add-on products, one should not forget that commercially successful add-on solutions in areas of high demand carry the danger that they will compete with systems developed later by ERP vendors; such packages come from a single source and consequently have fewer integration and maintenance problems than heterogeneous solutions.

Advantages of a packaged IS provided by a single supplier lie in the dovetailed architecture, reduced support costs, and lower costs per workplace. Because many different activities have to be coordinated in release changes, these may be time consuming even in this one-stop shopping case. Furthermore, there is the danger of becoming dependent on a single supplier of the IS.

In the end, many companies decide on a solution from a single vendor. An opinion poll conducted in spring 1999 among R/3 users in the production industries in German-speaking countries showed that about one-third of these companies had already decided to use SCM software; the Advanced Planner and Optimizer (APO) software from SAP has a significantly larger market share than its competitors. Surprisingly, as many as 87 percent of the companies expressed satisfaction with the solutions implemented (Dokaupil 1999). Another poll, also conducted in 1999, among 387 German companies showed that 68 percent of the companies preferred SCM products from ERP suppliers and only 21 percent SCM products from third parties (11 percent did not respond). The most important reasons for these preferences are simpler integration, avoidance of greater heterogeneity, and the solution competence of ERP vendors (NN 1999f).

In addition to software suppliers, several consultancies have also specialized in SCM (cf. section 6.4):

- Accenture, formerly known as Andersen Consulting, has formed the Accenture Supply Chain Ideas Exchange. A method "Supply Chain Value Assessment" is proposed [http://www.ascet.com/documents.asp?d_ID=285].
- IDS Scheer [http://www.ids-scheer.com/english/] emphasizes its contributions to the APO development with the resulting competitive advantage.
- IMG developed concepts for performance measurement before and after reorganization of the SC; it also provides a reference database for benchmarking [http://www.img.com/english/expertise/e_scma.htm].
- KPMG developed a five-stage SC model to enable organizations to manage and integrate the flow of information, money, and products within and beyond the enterprise; several strategies of SC partners can be coordinated by KPMG's e2eSM (enterprise-to-enterprise) Supply Chain Strategies approach [http://www.kpmg.com/search/index.asp?cid=241].
- PriceWaterhouseCoopers (PWC) has developed "Value-Based Management models" that help clients quantify SC improvement benefits. The SCM consultancy group is staffed with more than 3000 consultants worldwide. [http://www.pwcglobal.com/extweb/mcs.nsf/$$webpages/ supply%20chain%20management].

2 Application Systems in the Individual Business Functions

In this chapter we describe the major tasks that arise during order management in the most important functional areas. The SC "includes activities such as material sourcing, production scheduling, and the physical distribution system, backed up by the necessary information flows. Procurement, manufacturing, inventory management, warehousing, and transportation are typically considered part of the supply chain organization." But product development, demand forecasting, order entry, channel management, and customer service are also very relevant topics in SCM (cf. Bovet/Martha 2000, p. 17).

We shall concentrate on business and organizational changes associated with the SCM concept and on their effects on the intra- and inter-company IS. The sequence in which we describe the functional areas is based on the workflow in those industries that have special interest in applying SCM; these include the electronics, pharmaceutical, chemical, automotive, and aerospace industries.

2.1 Engineering

Engineering provides the technical concepts for product and process development. In a manufacturing company this includes, in particular, research and development, technical product planning, product design, development of process plans and programs for numerically controlled (NC) machines, and the technical preparations for layout, material flow, and quality assurance. Several other departments need the results of these engineering activities. Furthermore, outsourcing and cooperation with companies involved in engineering tasks are gaining in importance. From the SCM perspective, forwarding of product-definition data to systems used later within the same company's procedures (e.g., to develop NC programs, to provide BOM information for ERP systems, to feed data to CAM systems or automated quality control) and forwarding them to other enterprises are both highly relevant.

2.1.1 Development Teams

The mental barriers that exist between technical and commercial departments in many enterprises disturb the intra-company information flow and thus hamper market-oriented development activities. For this reason, mixed teams are recommended so as to obtain flows of specialized knowledge from different functional areas into the development work. It can sometimes be very useful to allow engineers direct customer contact (e.g., in a customer service department) so that they become more receptive to the wishes and complaints of the customers and are directly on the receiving end of suggestions for subsequent development work. Comparable experience can also be gained from assignments to other companies of the SC.

Mixed development teams often include employees of the SC partners with the aim of benefiting from their product-related knowledge and taking account of the possible effects on the material and information flows during product development rather than waiting until later stages; the reasons are similar to the rationale for incorporating external experts into value engineering teams. In the early phases of design projects, competing customers sometimes even cooperate with the same supplier. Once the principal solutions have been found, the detailed designs for company-specific products and processes may be organized in pairs. Because it becomes more difficult for a company that is cooperating in the SC to provide the know-how needed in the development process without contacting its partners, cooperation during the development phase enhances the stability of the SC and increases the switching costs if a company leaves it.

Integrating key suppliers, and possibly tier 2 up to tier n suppliers as well, early on in product development can have significant benefits. At the concept design stage, suppliers help to identify the most up-to-date technologies to be incorporated into manufacturing the new product. During the detailed design work the suppliers contribute to solutions for component and part development and are also involved in the selection of the most suitable materials and components. Suppliers may also assist in make or buy decisions. Infrastructural suppliers can define the most capable tooling, fixturing, and equipment. It is widely accepted that an early involvement of suppliers is beneficial to both buyers and suppliers. A prototype, WeBid, was developed to support cooperation in product development by providing

- a supply explorer, an SC model consistent with the new product development process,
- a bid explorer for inviting partners to bid for certain services,
- a partnership explorer for selecting a supplier, and
- a share explorer by which SC partners may share information of common interest (Huang/Mak 2000).

The following policies are regarded as key enablers for successful implementation of a technology SC (cf. Bozgodan et al. 1998):

- Establishment of integrated product teams
- Long-term commitment to suppliers
- Supplier-capability-enhancing investments
- Incentive mechanisms
- Joint responsibility for design and configuration control
- Collocation
- Target costing
- Database commonality
- Seamless information flow
- Retention of flexibility in defining system configuration.

Early integration of suppliers into product development has been studied under the auspices of the Lean Aerospace Initiative at MIT (Bozgodan et al. 1998). A survey looking at 40 military acquisition programs in the US defense aircraft industry found that early involvement depends heavily on the type of item being developed: Whereas in 75% of the programs the suppliers were engaged in the early development phases of major components, the respective numbers were 40% for subassemblies and only 20% for parts. Key suppliers with major subcontract responsibility are integrated in the early development phases, even before contract award, whereas suppliers of subassemblies and parts are often involved early for technical consultation on a variety of design issues but mostly selected in later stages of product development.

2.1.2 Concurrent Engineering and Simultaneous Engineering

Concurrent Engineering (CE) and Simultaneous Engineering (SE) are both designed to shorten the lead-times by paralleling, integrating, and standardizing the development tasks. Whereas CE is directed toward a distributed realization of certain development tasks, SE concentrates on developing products and processes in parallel.

There is a move towards the use of team-based SE within ship design in the US. The groups, labeled Integrated Product Process Design teams or Integrated Product Teams, bring together, in particular, representatives of the engineering, manufacturing, marketing, training, life-cycle support, operations, and purchasing functions and of suppliers in the earliest phases of design to consider all aspects of the ship's life-cycle concurrently. Many SE considerations can be resolved by developing a standard approach to design for vessels of a given class and by rationalizing the shipyard's production system prior to the elaboration of a specific design. The teams are usually co-located, but will also be able to communicate in virtual meetings via the Internet (cf. Parsons et al. 1999).

In CE, a development task is divided into subtasks; these are then solved in parallel, and the subsolutions are ultimately combined to provide the overall solution. Global Engineering attempts to make all 24 hours of the day almost completely productive, e.g., by having employees working on the same tasks in America, Asia, and Europe. In the near future, the subtasks may increasingly be solved by virtual cooperation, with the data being exchanged primarily via the Internet.

Product Change Collaboration (PC^2) systems are software solutions that support virtual engineering; their capabilities are

- a common source or database for full definition of the product the partners are trying to launch on the market,
- publishing services for sending the data to all partners as soon as changes are made, or even considered, to any of the relevant product data,
- the ability to analyze the impact of the proposed change and the ability to browse other relevant information about the affected products, parts, and documents, and also the change history for any related parts or products,
- the potential for collaborating with the other members of the team, proposing alternative solutions, commenting on the change or the implementation plan, and conducting electronic approval processes, and
- the ability to implement the change and to update the product configuration in the database. A key part of this is the ability to distribute the results to all other systems that need to be updated with the modifications of the product configuration (e.g., ERP and APS systems) (cf. Agile Software 1999).

Tools such as Agile Anywhere from Agile Software are used to manage the large number of engineering-change orders via the Internet. The PC^2 system allows every member of the product team to view what is happening to product designs and then take steps to plan accordingly. Every piece of information necessary to implement the manufacture of new products and to manage changes to existing products is readily accessible - BOM, part numbers, engineering-change orders, and drawings. The data in these systems can be fed directly into an ERP system (NN 1999e).

The frequently observed reduction of the vertical range of production often makes it desirable to pass some design responsibility to the suppliers. This reduces individual companies' own development ranges and results in some overlap of the development work between several companies. Telecooperation can bring about major changes in the development process.

Traditionally, information exchange was concentrated at the end of the product development phase. The production processes were planned on the basis of the completed product data. The (very ambitious) goals of SE are to shorten the development time, to lower costs, and to increase the quality of products and resources by paying attention to all product and process requirements of the responsible departments at an early stage (cf. Wildemann 1994, p. 32).

In addition to a partial parallelization of the development tasks, these can also be overlapped with the early phases of procurement or production logistics. In a poll, team meetings, training, jointly designed and used IS, increased use of methods, and joint releases were mentioned as being the most important measures for improving cooperation at the interface between engineering and procurement (Boutellier/Locker 1998, p. 119). In many types of companies, the interfaces between engineering and production attract particular interest (cf. section 2.1.4).

Both CE and SE demand a consistent database (as up-to-date as possible), which must be available to the cooperating units. Intranet and extranet technology and the accompanying access to databases can be used for this task. Recently, several papers have stressed the close connection between CE or SE, SCM, and the Internet (Kyratsis/Manson-Partridge 1999; Huang et al. 2000; Siemieniuch/Sinclair 2000). The theory of software agents (cf. section 2.3.4) has also been applied to CE (Sobolewski 1996).

Poor utilization and sharing of information may result in problems for an SCM-oriented CE (cf. Siemieniuch/Sinclair 2000, pp. 260):

- Incorrect descriptions of design issues, e.g., filtered by sales personnel
- Loss of integrity in the progress of the design
- Quality problems passed on from suppliers
- Differences in Computer Aided Design (CAD) systems, platforms, and conventions of use among SCM partners
- Incompatibility of systems and applications
- Difficulties in identifying parts
- Drawings not of satisfactory quality (or cannot be found)
- Inadequacy of geometry in drawings
- Working with out-of-date information
- Language difficulties when design data has to travel across language borders
- Reliance on face-to-face meetings.

Three-dimensional CE extends the focus from products and manufacturing to the concurrent design and development of capabilities chains: The design of the SC should be integrated with product and process development (Fine 2000, p. 218). From an organizational point of view, various approaches to implementing CE may be appropriate in different companies within an SC (Kyratis/Manson-Partridge 1999).

The protagonists of SE and CE scarcely consider that there are also good reasons for performing development tasks sequentially. The problems and the time required for the familiarization with work results provided by other persons put limits on the CE concept. It is seldom possible to structure the tasks in such a way that the described quasi-simultaneous processing can be achieved. Thus, interface problems may arise with CE even within the same phase of order management. The partial task overlapping requires additional coordination work, which does not occur in sequential processing. These possible risks and costs must be considered in the decision on whether or not CE should be made an element of SCM.

2.1.3 Variant Policies

Viewpoints on the appropriate range of variants often differ between the marketing and logistics functions. Marketing sometimes argues that the desired customer orientation means that every potential customer should be offered precisely the product variant desired. Technological progress makes it feasible in many (but not all) areas to produce a broad range of variants despite the resulting small lot sizes. Variants can be created by using similarity and configuration techniques (Schönsleben 2000, pp. 259).

A broad product range results in high logistics costs, which often cannot be recorded and allocated adequately by traditional accounting methods. The appropriate storage and maintenance of variant data in ERP systems requires significant effort; problems occur, for instance, during the data exchange in the SC when the partners use differently structured BOM (cf. Urban 1998). A very large number of variants means that not all of them can be explicitly documented. Demand forecasts of variants are much more difficult than those for product groups. Shortages may, for example, arise for variant U, although there are high inventory levels for a very similar variant X. ATP procedures (cf. section 3.1.4) are much more awkward to execute if many variants exist. These problems can increase in the upstream elements of the SC. ABC analyses show that although the majority of variants lead only to an insignificant proportion of revenues, they cause significant inventory and variant-dependent fixed costs. From this point of view, a "Design for Supply Chain Management" (Lee 1993) may also include reducing the number of variants and, thus, concentrating material flows, information flows, and financial flows. The "Design for Variety" method aims to help companies quantify the costs of providing variety and offers guidelines on how to develop products that incur minimum variety costs (Martin et al. 1998).

The following measures in the engineering work can influence the degree of internal diversity and associated logistics costs:

- Postponement strategies (Lee 1998) reflect the attempt to make products in such a way that the point at which variants are differentiated (the "Order Penetration Point") is reached as late as possible in the production process. Before this point, all work steps can be realized without reference to any specific customer orders, and the resulting preassembled modules may be stored; all subsequent work steps depend on specific orders from the customers and are not performed until the orders have already been received. Thus, postponement means "late customization". This goal can be achieved, for example, by standardizing parts and/or processes and by changing the sequences of work steps.
- Zero-based engineering investigates which parts would have been developed if all processes were reengineered. This can aid in recognition of unnecessary parts.

- Support for the designer in finding already defined parts, e.g., by Product Data Management (PDM) systems (cf. section 2.1.5).
- Promotion of "umbrella variants that provide a superset of features required by customers (Knolmayer 2001a).

Philips Consumer Electronics installs an integrated semiconductor in television sets that adheres to all standards that exist worldwide even though only one single form is relevant at a certain place of use. This key element is more expensive than parts that conform only to a national standard. However, from the perspective of an integrated SC, the solution with the part that can be used worldwide makes better economic sense (Handfield/Nichols 1999, p. 171).

Rather than solving every technical problem and offering every design extravagance, Schindler Aufzüge AG concentrates on a "super standard", the "ready-made elevator", a reliable, well-equipped elevator that satisfies all typical requirements (Schär 1999).

In R/3, the Product Variant Structure (PVS) System can be used to describe variants. PVS provides filter mechanisms to define function-specific views of product structures and descriptions (e.g., BOM for design, manufacturing, service or recycling). Every triangle in the display of the product structure (Figure 2.1) represents a part. Points placed next to the triangles indicate alternatives. The PVS is subject to R/3 engineering change management, which ensures a complete history of any changes made to product data. Rules can be formulated in PVS that permit R/3 to select one of several variants.

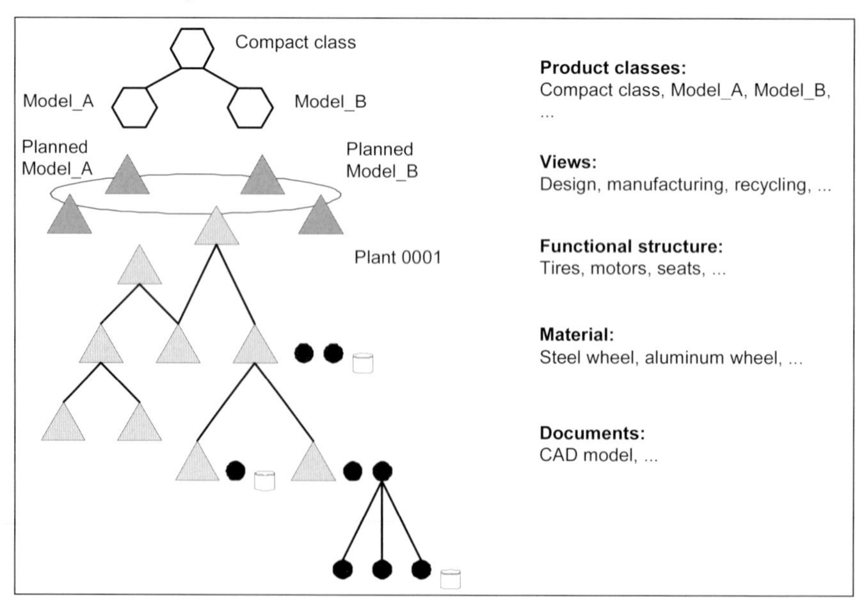

Figure 2.1: Visualization of the product variant structure in PVS

STEP (Standard for the Exchange of Product Model Data) processors can be used to exchange engineering data. Digital Mock-Up systems are available for visualization and simulations, e.g., for mechanical collision checks, electromagnetic analyses or feasibility studies. PVS can provide them with additional information using three-dimensional coordinates for parts (so-called transformation matrices) and meta-data from the document management system.

In addition to PVS, a module R/3 BOM also exists, which is recommended to be used for those parts that are not produced as variants. In particular, it is possible to use BOM to document manufacturing structures that lie ahead of the variant determination point and PVS for the remaining parts. The system performs integrity tests for this type of cooperation.

2.1.4 CAD Systems

The engineering department provides a precise description of a part in the form of drawings and BOM. Although only a small percentage of the total costs arises in this department, it is responsible for approximately 75% of the final costs. Development decisions are often made primarily from a technical perspective without precise knowledge of the current business data and without forecasts of these values for the product life cycles. This behavior comes about, to a certain degree, because the engineers have no or only limited or inconvenient access to the basic business data available in ERP systems.

In addition, the development department and the work planning department often do not have detailed knowledge of the implications of the various alternatives for manufacturing and logistics. This can result in decisions that prove to be inappropriate in the later phases of the life cycle. Many rejected development variants that could become relevant under different circumstances remain undocumented; this means that knowledge about the scope of feasible actions is lost.

CAD systems are often used in the development phase. These systems provide the following benefits:

- Simplification and acceleration of complex technical analyses, calculations, and creation of technical drawings
- Error reduction and quality improvements
- Department-specific views for product-defining data
- Fast access to earlier work results, which means easy exploitation of similarities and makes it possible to avoid having an unnecessarily wide range of parts
- Reduction of lead-times and of project efforts.

Computer-internal representations of the same object diverge significantly in different CAD systems. This may result in serious problems with data exchange if the SCM partners use different CAD systems. International efforts have led to open standards, such as IGES (Initial Graphics Exchange Specification) and STEP

(Standard for the Exchange of Product Model Data), and also to industry-specific standards (such as VDA-FS, a surface data interface standard originally proposed by the Association of the German Automotive Industry).

The Automotive Industry Action Group developed AutoSTEP (Phelps 1997). The focus of the project was on SC design tasks and communication processes between suppliers and customers. A STEP applicability matrix is proposed, in which the rows of the matrix represent the various functions that require product data and the columns capture such different levels of design responsibilities as

- commodity (catalog purchase),
- black-box design by supplier according to detailed specifications,
- gray-box design in which the supplier does primary design work with reference to some detailed specifications,
- collaborative design, and
- customer design, in which the primary function of the supplier is to provide the manufacturing capability.

The interfaces between CAD systems can be investigated by applying different types of tests. A transfer of product-defining data between SC partners assumes the use of the same CAD system (maybe even the same release of the system) or a commitment to conform to CAD standards. In the latter case, however, advanced non-standard functionality of a CAD system cannot be exploited, although this is often cited as a powerful argument in a particular system's favor in comparative evaluations of different software packages. The dependencies between technical and business data, as emphasized in CIM concepts, make the interfaces between CAD and ERP systems very important.

The R/3 System includes an open, non-industry-specific CAD interface. SAP has created a partner program that supports the coupling of various CAD systems to R/3 and certificates such interfaces. This work resulted in interfaces to leading CAD systems such as AutoCAD, CATIA, and Pro/ENGINEER. Nevertheless, these interfaces often need to be adapted to company-specific conditions.

One of the tasks of a CAD interface is to make the functionality of an ERP system available in the CAD environment. If R/3 is used, either the original masks can be invoked by a Remote Function Call (RFC) (cf. section 3.4.2) or the dialogs in the CAD system are reprogrammed to display only those details that are relevant for the designer, e.g., material master data.

An important aspect for chained task execution is that the earlier phases take account of the goals pursued in later phases. This should allow design solutions that provide the best possible support for manufacturing, maintenance, and recycling of the products; concepts such as "Design for Manufacture", "Design for Assembly", "Manufacturing by Design", "Design for Localization", "Design for Variety", "Design for Mass Customization", "Design for Postponement", "Design for Service", "Design for Environment", "Design for Recyclability", and "Design for SCM" illustrate these goals.

Hewlett-Packard attaches the power cable for printers in the warehouses rather than in the factories. This allows better responses to short-term, unexpected demands in different countries. To realize this concept, the design of the printers had to be changed to make the power supply module easily accessible from outside (Handfield/Nichols 1999, p. 169).

An SCM-conform design should allow controlling of the resulting material and information flows with limited effort. This can be achieved by reducing the number of parts, of manufacturing levels, and of process steps. Purchasing of preassembled subsystems rather than of individual parts (modular sourcing; cf. section 2.3.1) simplifies logistics, decreases quality problems, and results in reduced procurement, inventory, and transportation costs in the SC. New parts should be designed only if this is unavoidable. Consequently, CAD systems should contain or use classification techniques, to make it easier for designers to find documents about already existing parts. Hurdles to be overcome before new parts are created can basically take either of two forms:

- Bureaucratic solutions: A newly designed part may be used only after it has been released for use, e.g., by the company's standards office.
- Prices as a control mechanism: The development department is debited a certain amount if it defines a new part.

In the 1980s, Ford of Europe billed internally an "Administration Cost Barrier" of $ 25,000 for each additional part number (Baumhardt 1986, p. 809).

2.1.5 Product Data Management

Many of the CAD systems currently in use concentrate on the geometry of the objects and do not focus on organizational data. To counter this weakness, CAD and ERP systems are frequently combined with Product Data Management (PDM) systems. These support the complete, structured, and consistent management of all data and documents that need to be generated, processed, and forwarded in the company during the development of new or the modification of existing parts (data and document management). In addition, there are often functions to control and monitor the secure processing and forwarding of information. However, the main task of PDM systems is information linking, for example combining different databases, providing various application interfaces, and supporting different hardware platforms and operating systems (Stürken 2001). A PDM information center has been established at [http://www.pdmic.com/].

Consequently, such integrated solutions as are needed for a "Design for SCM" must contain interfaces not only between CAD and ERP systems, but also between CAD and PDM systems, and between PDM and ERP systems as well. The PDM system has to provide a comprehensive structure that coordinates the locally held data. This allows a user to access all elementary and aggregate data. For example, the PDM system can be utilized to display all CAD drawings associated with an order, determine the corresponding NC program for a certain CAD model or access purchasing documents to determine whether employees working on other projects have ordered identical parts. Web-based PDM solutions contribute to an effective SCM by supporting secure sharing of engineering data. Procedures for selecting a web-based PDM system have been proposed (Carroll 2000).

PDM systems strive to use data exchange standards. At least to some extent, this permits interfaces to be formed using STEP preprocessors and postprocessors. A PDM system may coordinate several heterogeneous CAD systems. Figure 2.2 illustrates the tasks supported by PDM systems in an SCM environment.

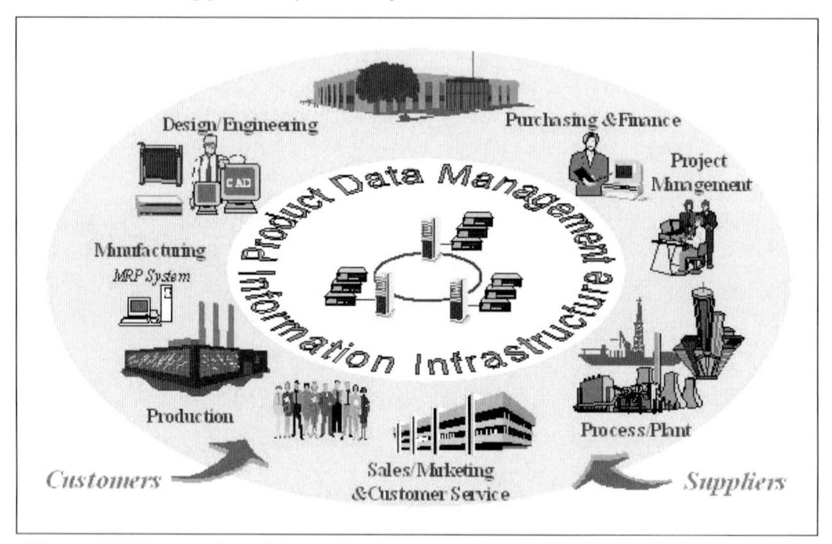

Figure 2.2: Product Data Management as central component of order management (CIMdata 1998)

A survey of a 1997 CIMdata conference reads as follows: "The greatest single issue facing the PDM industry is resolving overlaps and conflicts with ERP. Although there is no single correct answer for integrating PDM and ERP, the technical difficulties are being overcome; but the cultural barriers are still being breached. The PDM interfaces to the larger ERP software systems are usually two-way, with alternative methods for updating and a wide range of supporting technologies ... The concept of 'enterprisewide' PDM permeates much of the thinking of the PDM community, which means focusing on one of the key integration points between PDM and ERP systems - the bill of material ... A general consensus does seem to be taking shape in that respect: PDM should own the bill in the design and development phases of the product life cycle, ERP after release to manufacturing.

More and more PDM vendors are announcing interfaces with ERP systems. Not surprisingly, most interface activity is taking place with the R/3 System of SAP. There are about a dozen certified PDM interfaces now available, or under development" (Bourke 1997).

SAP offers PDM functions in its mySAP Product Lifecycle Management (PLM) solution (Eisert et al. 2000). PLM is delivered via mySAP Workplace (cf. section 4.1) and is designed as a platform for collaboration of all PDM users inside and outside the company. It contains functionalities, among others, for administration of documents, material master and BOM data, for product configuration and classification, and for change, workflow, and project management.

The PDM system can also operate stand-alone within an R/3 System architecture distributed over several computers and linked with R/3 using the ALE (Application Link Enabling) technology. A connection via the Internet is also possible. This allows the user to employ a web browser to search for R/3 documents, to display original design data, and to query product data.

A CIMdata report about collaborative PDM systems, published 2001-02-28, shows SAP as leader in this fast growing market [http://www.cimdata.com/PR02282001MOR.pdf].

Coordination is necessary if companies in the SC want to update the releases of their CAD or PDM systems and if the additional functionality used affects the data exchanged in the SC. Another important issue is configuration management: All parties involved in the SC should always know about the latest versions of the engineering documents.

For the future, changes in the task assignment between PDM and ERP systems are forecast (cf. Figure 2.3, where VPDM stands for Virtual Product Data Management). VPDM allows the engineer to study three-dimensional products in virtual reality, looking at some parts of the product and determining whether there are any inferences. The best suited design is handed on to the PDM system (Ptak/Schragenheim 2000, p. 175).

The increasingly used term "PDM II systems" also reflects the changes taking place in this market. In particular, it can be expected that a part of the existing PDM functionality will be provided in CAD systems and another (non-disjunctive) part in ERP systems. It is also assumed that VPDM will be integrated with project management systems to provide real-time and in-depth status for a design. Thus, a partner in the SC will be able to take a "virtual walkthrough" of the product by using a secure Internet access (Ptak/Schragenheim 2000, p. 184).

The role of PDM in extended enterprises is intensively discussed. The relevance of this topic shows that the cooperation of PDM systems is similar in importance to interfacing CAD systems. In general, SCM is not only concerned with business processes and data, but also with their engineering counterparts.

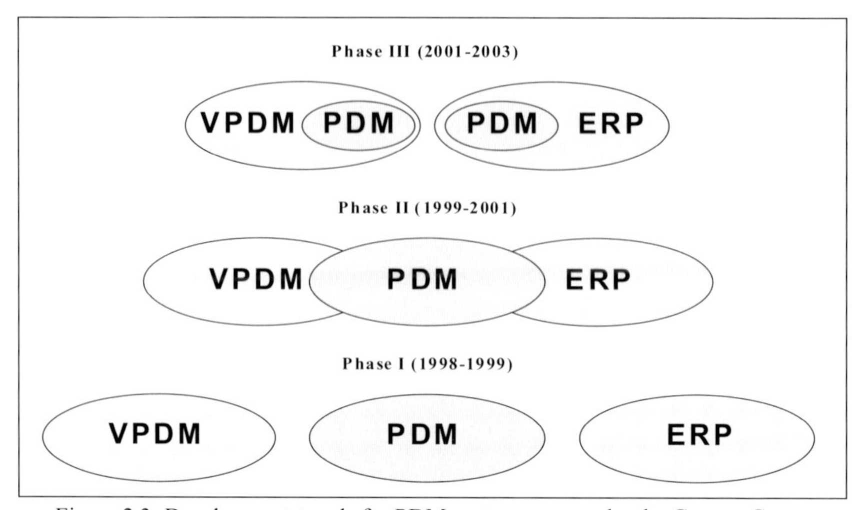

Figure 2.3: Development trends for PDM systems as seen by the Gartner Group

2.2 Sales

Although many SCM concepts are closely interconnected with logistics, SCM has also been considered from the viewpoints of marketing and sales. In this section we consider current developments in sales activities and their impact on SCM; many of these trends are closely related to new technical capabilities of IS.

Basu/Chandra (1996) argue that the key factors for improving SC behavior can be found at the interface between marketing and production. They regard sales promotion as a main factor driving the demand amplification; this policy has therefore also been termed as the "dumbest marketing ploy ever" (Sellers 1992).

2.2.1 Approaching the Customer

Database marketing

The aim of Database Marketing (DBM) is to improve the quality of sales activities and to reduce the costs of marketing campaigns. Targeted contacts (e.g., conventional direct mailings, fax or E-mail) should reduce the number of expensive customer visits.

The following goals are pursued through DBM:

1. *Individualization.* DBM is used to determine a product quotation based on the individual preferences of the recipient. The customer approached should receive an adequately individualized offering selected from the total product program and from possible conditions of delivery.
2. *Clustering of the customers and prospective customers in widely homogeneous areas.* For DBM in the consumer goods area, micro-geographical segmentation is an important analysis instrument: Geographical references (e.g., name of the city or its zip code) and sales behavior-oriented information (e.g., purchasing power and household situation) result in regional types within which the customers are widely homogeneous with regard to their purchasing behavior and potentials.
3. *Improved measurement of outcomes of customer contacts.*
4. *Early recognition of sales potentials.* Companies have the ability to recognize early purchasing signals and to offer the appropriate products to prospective customers. For example, a furniture store could apply purchase event chains (such as building permission applications => approvals => house construction => furnishing) to forecast what products a household will buy in the future.

DBM requires a carefully structured and maintained customer database. In addition to basic data (such as address, date of birth, gender), it should also contain information about individual preferences, sales promotion data (customer-specific marketing measures), and response data (customer reaction to sales promotions). Additional internal and external information, e.g., about payment history (from accounts receivable systems) and market research data, are also useful for DBM. The companies cooperating in an SC may merge and restructure their single databases, forming large data warehouses to support the extended enterprise.

Some customers do not wish detailed profiles about their purchasing behavior to be held on the supplier's computer. This may be especially relevant if other members of the SC can access this information and combine it with their own data about this customer. A balance must be struck between the overall benefits of the more precise approach to the customers and the possibility of annoying some of them.

Data mining

Data mining covers various procedures for the (at least partially) automated analysis of large, structured volumes of numeric data. It attempts to identify patterns, to recognize differences between groups of data records, and to determine their characteristic attributes. Typical insights obtained are that customers with specific combinations of properties prefer certain products or groups of products. Such knowledge must be provided with details about its level of reliability and presented in an easily understandable form. Methods from different areas are applied for data mining; these include database research, statistics, machine learning, fuzzy-set theory, neuronal networks, and genetic algorithms.

The SD-CAS component of the R/3 System serves the sales and marketing departments as an instrument for the management of customer contacts and acquisitions. An E-mail may contain a message (invitation to a trade show, information about a new product, etc.), annotated links to a web-page, and attachments (e.g., white papers, brochures, product descriptions). Depending on the target group for the promotion, the list may consist of addresses of existing and prospective customers. Background information about these organizations and groups of people is stored as master data. Specific data, e.g., earlier correspondence, are stored under "contacts". The "Lead Tracking" function can be used to display the document flow for a mailing action to determine whether follow-up contacts or sales have resulted.

SAP is offering solutions for one-to-one marketing, database marketing, and data mining concepts as part of its mySAP Customer Relationship Management (CRM) applications. Data mining can be realized with reference to the BW and the integration of an add-on product (e.g., Intelligent Miner from IBM, DeltaMiner from Bissantz & Co). Typical features of the DeltaMiner tool are

- *Analysis Chain Technique*
- *Integration of statistical, business intelligent and data mining methods*
- *Cockpit features*
- *High level of automation (routine queries are completely delegated to the computer)*
- *Support by an Analysis Wizard which "observes" the user and checks the data for remarkable findings*
- *Integrated data quality analysis*
- *Sound and graphical animations enhance the perception capabilities of the user*
- *Speech output*
- *Automatic Quick-Configuration [http://www.bissantz.de/].*

2.2.2 Quotation System

Recording a customer inquiry and an associated quotation in an IS is particularly desirable if a large number of customer requests and quotations need to be processed or if product specifications and quotations can be derived from largely standardized elements (Mertens 2000, pp. 50). In the latter case, the input of a specification allows the program to determine the parts of tables or databases that are relevant, to combine them using specified rules, and to calculate the product defined in this way.

Because developing quotations for customized production requires significant effort, the use of resources of the involved functional areas, e.g., the development department, must be planned very carefully. To obtain a basis for selecting inquiries and determining what constitutes adequate effort for providing a quotation, information must be gathered about the responsiveness of customers to earlier quotations. Some companies maintain checklists that must be processed on a computer system before creating a quotation (cf. Figure 2.4).

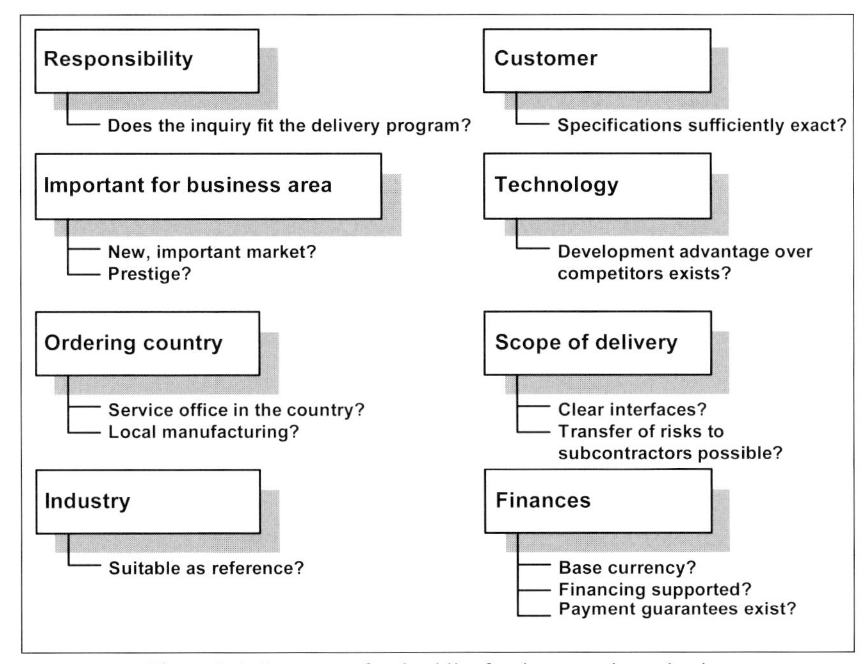

Figure 2.4: Structure of a checklist for the quotation selection

A modern sales medium is the electronic product catalog, which supplements or replaces traditional paper catalogs. It exists in many different forms, being accessible, for example, from so-called kiosk terminals set up at locations subject to heavy pedestrian traffic or available on CD-ROM or on web pages. We will discuss these catalogs in conjunction with the procurement function (section 2.3.4).

Know-how databases are important means of knowledge management. They are used to store the solutions that a supplier has designed for its customers in the past and thus save the associated know-how of the development task. When new customer-specific quotations are prepared (at least some) of the experience gained in earlier contacts with this customer can be drawn on. For multinationally operating companies in particular, this procedure can help to prevent application knowledge from becoming too scattered within the company. More specifically, a database may contain CAD data with detailed explanations, extracts from trade journals, test results, product-specific customer suggestions, complaints, maintenance data, and reasons why quotations did not result in orders (lost-order information).

Use of know-how databases

The information is processed using modules for the acquisition, classification, and retrieval of solution descriptions. A sales employee or a designer enters the data at a decentral work place when the first samples are developed or released for a customer, or when a quotation has been prepared. Employees in all departments and locations can search the database to find existing or similar problem solutions. To support the information retrieval, the documents may be classified by descriptors. The computer assigns some data (e.g., document number, date, creator, department) automatically during the acquisition; other information, such as the data of the customer and the sector of industry in which the customer operates, have to be manually entered or selected using a menu system. Automatic generation of descriptors from the document after eliminaton of filler words is conceivable. A member of the staff can then accept, change or extend the automatically created descriptors. During retrieval, the system searches for all documents to which the query parameters have been attached. A full-text retrieval of the information stored should always be provided.

SAP does not supply an independent component for know-how databases. However, such a knowledge base can be built up via the R/3 document management tool with its information retrieval capabilities. A full index may be constructed by using indexing tools such as AltaVista or Verity, which can be connected to R/3 [http://www.altavista.com/; http://www.verity.com/]. The PDM component of R/3 supports the management of product definition data (cf. section 2.1.5).

Configuration

The sale of system solutions (cf. modular sourcing, section 2.3.1) and a trend toward mass customization make the product configuration a significant part of the quotation process. High configuration costs and a low percentage of inquiries resulting in orders make computer support for this task essential. Configuration engines enable complex products to be determined from standardized components and stored rules in accordance with the customer's specification. A configuration check assures that the customer is offered technically consistent and, whenever possible, appropriate solutions from the start.

Figure 2.5 illustrates various types of configurations. In the case of an active configuration, the IS determines the product proposal. It either offers its suggestion automatically and in total or successively recommends partial solutions that the user can accept or reject. With a passive configuration, the employee configures and the machine offers criticism either after each decision or when the product is finally designed. This procedure is known as a "criticism system" (Mertens 1994b).

The Make-to-Order function of SAP R/3 permits all four configuration variants.

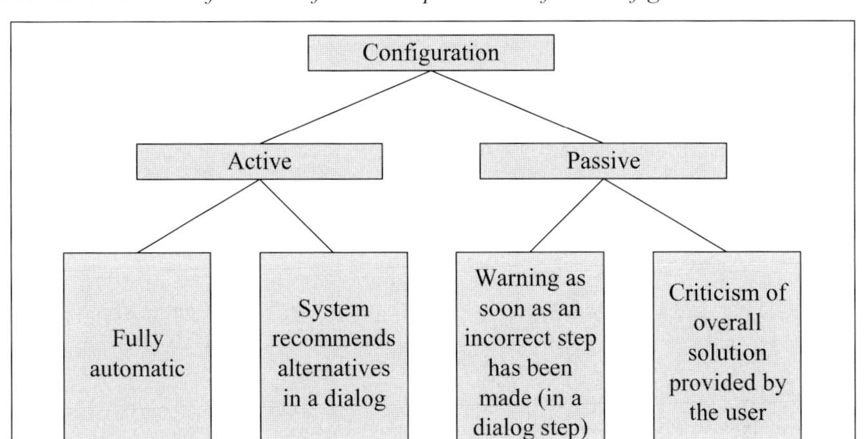

Figure 2.5: Types of configuration

In a manner analogous to the sharing and merging of marketing databases (cf. section 2.2.1), the partners in an SC may also pool their technical knowledge to give an SC-wide common knowledge management that serves the individual quotation systems by way of sophisticated know-how databases and configurators.

Target costing

Sometimes, manufacturers and their engineers develop technically advanced products with a broad range of functions and superior characteristics even if the customer does not need them or is not willing to pay for them. Customers often define a limited budget for a specific procurement. Target costing deals with this situation.

From a conventional point of view, the costs for a product generally result from the set of its features and functions; these costs form a basis for pricing. In target costing, the costs of the product are derived from the maximum price that a customer is willing to pay. The specification of the product features follows in a second step, in which this price limit is kept in mind. Instead of aggregating bottom-up individual cost elements to obtain a starting point for pricing, the supplier proceeds top-down and determines which costs are permitted without exceeding the price limit. This approach may proceed as follows:

First, find out what features and price customers actually envisage for this product. Market research instruments can be used for this task. Sales experts, or even a specific customer who is a potential buyer of the product, may contribute valuable information about appropriate prices. The result of this phase is an exogenously influenced target price.

Secondly, subtraction of the desired minimum profit from this target price gives the target costs or "allowable costs". The manufacturer has to investigate what costs result for production of the planned product. Because these "drifting costs" will often exceed the allowable costs, ways of reducing the costs of the new product must be sought. The target price may be approached by looking for compromises between nice-to-have features and the associated costs of adding or removing components. The configuration and calculation processes must be linked for evaluation of the effects of any particular step. Depending on the product, there may also be complex rules for the configuration that have "hidden" cost-relevant implications, e.g., if the installation of a component required by a customer means another part is necessary or if the removal of one part makes other components obsolete.

Reducing the drifting costs will often require intra- or inter-enterprise cooperation in the SC; in the latter case the following mechanisms can be used to foster inter-organizational cooperation (Cooper/Slagmulder 1999, pp. 143, p. 161):

- Functionality-price-quality trade-offs
- Interorganizational cost investigations
- Concurrent cost management programs.

The buyer may specify individual target costs for each component delivered by a supplier or it may bundle the target costs, giving the supplier some freedom to identify the points where it may concentrate its cost reduction efforts. Target costing may also be used as a measure in supplier development; innovations by the suppliers may be rewarded by allowing higher target costs (cf. Cooper/Slagmulder 1999, pp. 194). Empirical results (including 11 case studies) on the relationships between SCM and target costing are given by Ellram (1999).

Target costing should be applied with caution, because, in a similar way to value analysis, it involves the danger that a wide range of variants will result that can no longer be handled economically.

Target costing can be realized with the R/3 Project System (PS). Based on detailed work breakdown structure elements that may relate to BOM, budget limits can be defined; the system automatically generates messages if these thresholds are exceeded. The budget-checking procedure can be customized according to the relevant business rules.

Mass customization

Electronic sales aids are an important assistance for mass customization. This concept tries to combine the principles of mass production with those of make-to-order production. The prospective customer can specify features of the product according to its wishes. The customer configures its product from a wider range of variants or components. It is necessary to ensure that a "do-able" product results. The customer specifications are usually sent via telecommunications (e.g., the

Internet) to the factory and initiate an assembly process (assemble-to-order). The IS of the SC partners have to make sure that components ordered from third parties arrive at the factory in time. A mass customization system attempts to make use of the information gained during the interaction between business partners to build a permanent customer relationship from which both parties should profit. However, additional costs are usually resulting from making a process mass-customizable and, therefore, the problem of determining the optimal extent of mass customization arises (Knolmayer 2001a).

SAP provides a solution "Order Entry with Configurator" using the Internet Applications Components (IAC). The data are entered directly into the SAP R/3 System via the Internet; the production order is then scheduled immediately.

2.2.3 Order Entry and Verification

Order entry

The goal of this activity is an economical collection of external data for integrated information processing. In addition to conventional methods of order entry there are solutions that use the Internet for this task.

SAP systems offer a wide range of technical options for modern order entry. In particular, the EDI transmission based on Intermediate Documents (IDocs) enables ordering information to be transferred from an external system to the R/3 SD (Sales and Distribution) module. When two R/3 Systems communicate with each other, ALE should be used for interfacing. If the SAP Internet Transaction Server (ITS) and the Internet Application Components are used, the orders received from the customer can be entered into the R/3 System via the Internet. As part of its mySAP CRM application, SAP is currently developing additional components that permit sales staff working in the field to prepare the complete order entry (e.g., with exact product configuration, price calculation with customer-dependent discounts, and delivery date) "offline" on a notebook (cf. section 2.2.4; Angeli et al. 2001).

Order verification

An order must be subjected to detailed integrity checks before it is accepted and stored (Mertens 2000, pp. 67). Such checks can be omitted if the order applies to a recent quotation for which the checks were performed during the quotation phase and if it can be assumed that no significant changes have occurred subsequently.

Technical validation

If the product to be supplied is based on customer-specific wishes, it must be clarified whether it can be manufactured according to this specification. This validation covers factors such as unusual dimensions, tolerances, performance characteristics or unusual combinations of properties. Especially for companies with rapidly changing products, it is necessary to ensure that the model ordered is not obsolete. Because of environmental regulations, e.g., in the chemical industry, it may be necessary to test the product for prohibited substances. The cooperation in an SC may provide access to special knowledge of partners which is not commonly available.

The R/3 configuration mechanisms allow technical validations. To avoid the use of combinations of substances that the environmental regulations do not allow, a special query can be triggered by the master record.

Credit standing review

A limit for the sum of the open receivables and the value of the orders that have not yet been invoiced may be stored in the customer master data. However, because the customer could have significantly reduced its credit need by the time of delivery, this static consideration often does not suffice and a dynamic validation is needed. The accounts receivable program calculates the time between the invoicing date and the actual payment date for each payment. This time is smoothened and the expected value determined in this way is stored as the typical time to payment in the customer's master data. It is used by the credit standing module to forecast the dates of payment arrivals. This makes it possible to estimate whether credit limits might be violated on the occasion of the order just being checked. If the customer has sometimes not paid within some specified period in the past, the company may require that the person responsible be informed before confirmation of the order. If partners in the SC have common customers they may exchange their information on the past payment behavior and the current debts.

The "credit management" functionality in the R/3 sales (SD) and finance (FI) modules covers this area. An upper limit is defined in the customer master data for the static consideration. Rules that also take account of the current payments can be specified for dynamic checking. The customers are assigned to risk classes for which a differentiation is made, for example, between subsidiaries for which there is minimum payment risk and new customers with a high danger of loss. Depending on the class they are assigned to, the credit worthiness of the customers is handled differently. In addition, different methods of payment can also be defined. When payment is to be made with a credit card, the amount is authorized by the clearing-house of the credit card company and the transaction is handled by the R/3 System.

Delivery date check

A delivery date-checking module can be used during order entry to forecast whether the customer's favored delivery date can be met. This check first requires that the inventory levels be queried; the search starts with the end products, continues with the intermediate products, and ends with parts purchased from suppliers. If it is possible to deliver from the finished goods warehouse no further date checking is necessary. If this does not apply, the operations needed to manufacture the end products can be planned with due consideration for any available inventories of intermediate products. In the extreme case of a make-to-order product, all operations starting with the first operation on raw materials must be scheduled. If necessary, the time of procuring supplier parts must also be considered. Strictly speaking, the complete procurement and manufacturing process, including, e.g., all batch size changes resulting from the order to be checked, would need to be scheduled before it was possible to determine whether the desired customer delivery date could be met. This would obviously be a very demanding task.

In SC the tasks described above are even more complex. Software vendors have developed special SCM modules, usually called ATP (Available-to-Promise), for supporting delivery date decisions (cf. section 3.1.4).

R/3 permits the "planned delivery time" to be calculated when a product becomes available; this takes into consideration all available stock, future deliveries, capacities for in-house production, required resources, and the time required to obtain parts from suppliers. The most powerful and highly differentiating SAP offering for date checking is the ATP module of the APO system (cf. section 3.1.4).

2.2.4 High-Level Sales Strategies

Electronic Commerce

Three categories of EC are often distinguished:

- Business-to-Business (B2B)
- Business-to-Consumer (B2C)
- Business-to-Government (B2G).

A company practices B2B if it uses network services to transfer its orders to the suppliers and to receive orders from its customers, to receive or send invoices, and to process its payments. This type of EC proved itself many years ago, in particular with the use of EDI. However, with the arrival of the Internet, the B2B concept allows the establishment of electronic markets that often are organized vertically for single industries.

SAP offers a Business-to-Business Procurement solution (cf. section 3.2). This product permits business relationships based on Internet technology. It makes it possible to control and handle the procurement process, including the inquiry, status queries, issue of an invoice, and management of the payments.

The B2C category is also called electronic retailing and came up with the broad acceptance of the Internet by consumers. The Internet provides many online stores and shopping malls that offer consumer goods, including selected durables, such as computers and vehicles. When the manufacturer allows direct purchasing, intermediaries may lose their traditional role in the distribution channels. The physical distribution, especially of cheap products, is often a major cost component in B2C relationships. In addition, consumers sometimes repudiate credit card payments after delivery of the goods. Thus, additional mechanisms for secure trading, such as a Secure Socket Layer and digital signatures, are necessary for management of B2C relationships.

The R/3 Online Store component can be used to create an electronic product catalog on the Internet. A customer may use a web browser to access it and to query, configure, and order products. The data entered are transferred in HTML format to an SAP Internet Transaction Server which transports them to the R/3 System where they are processed.

All partners in an SC also have different types of contacts to government authorities. The third category of EC, B2G, takes account of the special properties of these contacts. In the US, for example, details of future government procurement programs are published on the Internet and suppliers have the option of responding electronically. Although this category is just starting to develop, it may rapidly gain in importance if governments decide to stimulate EC via B2G.

SAP is working on a development project for B2G.

EC ignores geographical and national boundaries. Because the important networks operate globally, EC enables even small providers to have a global presence and to operate worldwide. Customers can choose a specific product or service from many providers. Generally recognized standards will develop that simplify the business processes between providers and customers. Brand names create trust and will gain importance. The selection of a partner will increasingly depend on its capabilities to handle electronic business or even to cooperate in SC structures. Global EC is hampered by language barriers, the hardening of price differentiation, the sometimes unclear legal environment, and the difficulties of providing after-sales services on site.

Efficient Consumer Response

Efficient Consumer Response (ECR) is a bundle of methods which, as an element in a trust-based cooperation between manufacturers and retailers, aims at removing inefficiencies along the value chain. The goal is to provide a benefit to all the parties involved that they could not achieve by themselves. The American Food Marketing Institute (FMI 2000) suggests four major strategies within ECR:

1. Efficient Store Assortment - addresses how many items to carry in one category, what type of items and in what sizes/flavors/packages, and how much space to give to each item. This is closely linked to category management.
2. Efficient Replenishment - focuses on shortening the order cycle and eliminating costs in it, starting with accurate point-of-sale data. It includes efficiencies

to be gained by using continuous replenishment programs, EDI, cross docking, computer-assisted ordering, and new receiving techniques.

3. Efficient Promotion - addresses inefficient promotional practices that tend to inflate inventories and practices whose effects may not be fully passed on to consumers to influence their purchase decisions.
4. Efficient New Product Introduction - addresses improvements to the entire process of introducing new products, which is subject to high failure rates, thereby bringing extra costs into the system.

"Category Management" considers groups of goods rather than individual products as units. A category manager is responsible for a certain group of products. He can, for example, exploit a particularly low-priced article, A, using it to stimulate sales of another product, B: thus, A subsidizes B but the profit of the product group may increase.

In the "extended enterprise", category management can also be applied to the logistical processes in the SC. Category management is then the basis of the agreements between suppliers, purchasers, and logistical service providers. For example, the critical success factors, the associated management information systems, the accompanying data and, possibly, also the determination of storage locations in warehouses are slanted to the "categories".

The SAP Retail module IS-R delivers functionality for ECR and Category Management (SAP Category Manager Workbench). The APO system provides tools for VMI, a concept that consumer goods manufacturers are already using intensively (cf. section 5.1.1).

Quick Response

Quick Response systems are models for partnerships of manufacturers and retailers that aim at accelerating the information and product streams through the use of common IS. Quick Response is a form of Just-in-Time (JiT) delivery systems and is used, in particular, in the textile industry (Blackburn 1991). The following example illustrates this procedure: The clothing retailer's computer is connected to the IS of a ready-made clothing manufacturer. This enables the retailer to see what free capacity the manufacturer has in terms of storage or production. The retailer can place orders quickly and be relatively sure that the products have not been sold to another retailer in the meantime. Because retailing companies increasingly use merchandise information systems, quick response systems could soon become an interesting feature of an SCM solution (Mertens 2000, p. 109). The use of scanners, barcodes, and EDI supports the capability to respond immediately to any actions of the customers.

Some apparel and textile manufacturers are skeptical about the benefits of Quick Response systems because they assume that these will shift inventories upstream. Technical solutions and realignment of contractual relationships may be helpful to avoid this effect and to improve the overall efficiency of the system (Lee/Whang 1999, p. 638).

The R/3 System provides the functionality needed to realize the Quick Response concept. Contact can be made by way of a Web browser. SAP provides Internet Application Components that permit Internet access to an R/3 System via the Internet Transaction Server (ITS).

Customer Relationship Management

The main issues in Customer Relationship Management (CRM) are:

- IS provide standardized support for the customer relationships for the marketing, sales, and after-sales (service) cycle, where boundaries between these phases decline in importance.
- In large enterprises in which many people maintain relationships with an individual customer, efforts are devoted to presenting a uniform image to the customer ("one face to the customer").
- All companies involved in the "extended enterprise" have access to information on their common customers.

One core idea of conventional CRM systems is that diversified enterprises or even groups present the above-mentioned "one face to the customer". This concept may also be transferred to the SC; in this case, the partners would have to organize common databases (cf. section 2.2.1) making relevant attributes of the customers accessible within the SC.

The mySAP CRM Initiative deals with the following concepts:

- *One-to-one marketing (cf. section 2.2.1)*
- *Storing profiles of contact partners at the customer with the aim of sending them special messages ("profile specific addressing"); this provides the basis for database marketing (cf. section 2.2.1)*
- *Support by providing product configuration tools (cf. section 2.2.2)*
- *Mass customization (cf. section 2.2.2)*
- *Provision of current messages ("latest information/automatic news update").*

As part of its mySAP CRM solution, SAP offers additional tools to support field staff (mobile sales force). To prepare for customer contact, the salesperson can create, retrieve, and process, for example, information on the customer (requests, orders, goods and services, offered or provided, facilities installed at the customer's site, notes about creditworthiness, opening hours, etc.) and receive other information, for instance on products or special offers, from the server. The article display is provided by a product catalog (cf. section 2.3.4), which also allows multimedia elements (e.g., video sequences). The "Configuration Engine" permits configuration of products on a notebook. The knowledge-based system contained in R/3 is also available offline on the notebook. The salesperson in the field can access not only operative data but also transfer extracts from the Business Information Warehouse to his notebook and then process this information with the analysis and reporting tools available on it. The notebook has its own database, in which, e.g., orders, products, a calendar, and contacts can be maintained.

In addition to laptops, a personalized digital assistant or a mobile phone can also be used as a terminal to which the information is transferred.

In summary, the use of new technology allows the distribution of information to salespersons and provides them with easy access to information that is available only in central systems. This permits more intense information exchange with customers and new methods of data access and data sharing within an SC.

2.3 Procurement

The term SCM focuses on the supply side of business activities and thus on the procurement function. The cost of procurement as a percentage of total costs rises with weakening of vertical integration. The percentages listed in Table 2.1 indicate that success potentials can be found increasingly in the procurement function. One of the main success factors in SCM is involving the suppliers as partners.

Table 2.1: Typical percentages of purchasing costs for various types of industries (Thru-put Technologies 2001)

Manufacture of	Purchasing	Production	Distribution
non-durable consumer goods	30 - 50%	5 - 10%	30 - 50%
durable consumer goods	60 - 70%	10 - 15%	10 - 25%
industrial goods	30 - 50%	30 - 50%	5 - 10%

For the manufacturing industries only about 40% of procurement spending originates from buying production materials, and about 60% from indirect goods and services. Indirect goods are, for example, Maintenance, Repair, and Operations (MRO) goods, capital equipment, computers and software, magazines and books, office equipment and supplies, real estate, and vehicle fleet (Segev/Gebauer 1998).

The importance of the purchasing function for SCM is emphasized by the Purchasing And Supply-chain Benchmarking Association [http://www.pasba.com/]. This group conducts benchmarking studies to identify practices that improve the overall operations of its over 1100 members (as of 2001-05). Objectives of PASBA are to

- conduct benchmarking studies of important procurement and SC processes,
- create a cooperative environment in which full understanding of the performance and enablers of "best in class" procurement and SC processes can be obtained and shared at reasonable cost,
- obtain data on process performance and related best practices,
- support the use of benchmarking to facilitate procurement and SC process improvement and the achievement of accuracy, timeliness, and efficiency.

2.3.1 Sourcing Strategies

Global Sourcing

The globalization of business relationships stimulates the procurement especially of raw materials from worldwide sources ("global sourcing"). Goals of global sourcing are, among others, alliances with so-called world-class suppliers, market entry in countries with protectionist business policies, realization of saving potentials, managing currency risks, and provision of benchmarking information (cf. Boutellier/Locker 1998, pp. 134). The Internet improves market transparency and facilitates global sourcing strategies.

A case study on global sourcing at Mercedes-Benz Argentina shows the complex services performed by the forwarding agents (in this case, the Bremer Lagerhausgesellschaft). The process of global sourcing implemented by Mercedes-Benz Argentina is taking place in eight steps (Faber 1998, pp. 288):

1. *Preliminary transport*
2. *Consolidation*
3. *Transport to the harbor*
4. *Sea and air freight*
5. *Customs clearance*
6. *Container transport to the delivery point and container handling*
7. *Deconsolidation, warehousing, and supply to conveyor belts*
8. *Return transport of reusable containers.*

Single Sourcing

Traditionally, companies have bought important parts from several suppliers ("multiple sourcing"). Reasons for this strategy are competitive advantages, avoidance of dependencies, and risk management.

The concepts of procurement logistics and SCM demand close cooperation with the suppliers to permit, e.g., JiT deliveries and the transfer of quality control to the suppliers. This need usually results in a reduced number of suppliers for a specific part and, in the extreme case, in a single sourcing strategy. Advantages of single over multiple sourcing lie in cost degression effects and in the necessity for establishing a trustful relationship with the preferred partner. The selection of the supplier is of particular importance for single sourcing strategies. Guidelines for consolidating the supply base are available (Riggs/Robbins 1998, pp. 152).

Concentrating procurement on a few or even just a single supplier results in mutual dependencies. On the one hand, the "Lopez effect", seen at GM and Volkswagen, describes the market power of the buyer, which, in particular, may result in vehement demands for price reductions. If a supplier does not accept such requests, the customer can threaten to change to a different supplier, and can actually do this. For suppliers subject to a cluster risk, i.e., high dependency on a single customer, this may threaten the company's existence.

On the other hand, as seen in 1998 in the relationship between Ford Europe and its door lock supplier Kiekert, for example, troubles with a single supplier can bring production to a standstill. At least in the short run, alternative deliveries are impossible in such circumstances, because there are no cooperations with other suppliers that would enable any alternative (even stopgap) arrangements to be made. After this experience, Ford Europe announced that it would be reconsidering its single sourcing policy.

Modular Sourcing

The number of manufacturing levels is often reduced for logistical reasons. Consequently, compared with the procurement of individual parts, the acquisition of complete assemblies from system suppliers ("modular sourcing") is becoming more important. This may have the following consequences in the corresponding functional areas (cf. Weiß 1996, p. 134):

1. Engineering
   - Using special knowledge of the suppliers
   - Rapid implementation of technical progress
   - Reduction of development times
2. Procurement and materials management
   - Reduced number of suppliers
   - Fewer transport activities, reduced logistics costs
   - Lower transaction costs
   - Smaller quality control effort on receipt of goods
   - Reduction of inventory levels
3. Production
   - Reduced complexity of production planning and scheduling
   - Better control of time and quality of the processes
4. Administration
   - Reduced administrative tasks
   - Less demand for scarce management capacity.

Just-in-Time Procurement

In a similar manner to how the JiT principle is used within companies, it can also be applied between companies to organize the material flow on the basis of the pull principle. In particular, companies that are active in the automotive industry ask that their most important suppliers coordinate their deliveries closely with demand of the automakers. The parts are not delivered to a traditional arrival area but immediately to the precise location where they will be required, e.g., to a specific workstation along the conveyor belt. Such production-synchronous procurement implies close cooperation of the automobile manufacturer with the supplier and/or the forwarding agent.

Despite the importance of modern forms of telecommunications, the geographical vicinity of companies cooperating in an SC does have advantages for example with regard to organizational learning (cf. Dyer 1996, p. 289, with notes on the Japanese automotive industry and on Silicon Valley), but also for product development and establishment of JiT systems. Because of the imponderabilities of delivery times that result from long transport distances, production-synchronous procurement is based on the assumption that the quantities to be delivered are already physically close to the recipient and only a short transport distance needs to be covered once a request for the goods has been received. Consequently, some suppliers build factories, or at least warehouses, close to the customer’s premises. Because of the resulting inefficiencies, regional shipping companies offer to build warehouses near large customers to which the suppliers send their goods; the customer can release orders synchronously with its progress in production activities.

Information on the current demand patterns of the customers is very important for the suppliers. IS are not needed if a traditional kanban system is used to control deliveries (cf. section 2.4.3). However, because information on a customer's inventory levels gives the supplier an idea of the demand pattern and thus provides a better basis for planning, information and communication systems are ususally needed for coordination of the JiT relationships in the SC. These require close matching of the business processes of the companies involved and are thus primarily realized for the high-value A-parts. In this sense, JiT is closely related to the SCM concept.

The aim of JiT deliveries is to avoid manipulations of the parts during receipt of goods with consequent lowering of the inventory levels. The quality inspection may be performed according to mutually agreed rules at the supplier's location; if necessary, such quality control can also be performed by a third-party and/or at a neutral location (e.g., in a warehouse that a regional carrier operates near the customer’s plant). The company receiving the goods can determine the measures of quality control and also decide whether it will be involved in the inspections and, if so, to what extent. The transfer of the quality inspection is a typical example of how the SC cooperation changes process chains.

2.3.2 Centralized and/or Decentralized Procurement

The purchasing function may be organized in a centralized or a decentralized manner or in a combined form. For combined procedures, threshold values that initiate a centralized procurement can be specified if they are exceeded in decentralized purchasing. A current trend is direct purchasing via the Internet by user department employees. Especially for C-materials, tendencies to decentralized purchasing can be observed (cf. section 3.2). In this case, staff members can independently choose and order materials within a customized and restricted range that has been authorized (budget limits, range of suppliers). A web-based desktop purchasing system provides information on potential sources and their terms of delivery and payment. A variant of this policy permits decentralized procurement if the prices of decentrally purchased goods do not exceed those offered by the central buying department by more than some specified percentage, for example, 3%.

Alcatel SEL started the "Buy Direct" project in January 1998. The concept allows for direct ordering by the department that needs the goods and for electronic invoicing; this results in reduced loads on the purchasing, controlling, and accounting departments (Schäffer et al. 1999).

Recently, with the so-called material group (or commodity) management teams, attempts have been made to combine the advantages of centralized and decentralized procurement by using an inter-organizational coordination forum. Interdisciplinary teams specify sourcing policies, analyze the procurement processes, conclude company-wide buying agreements, prepare outline agreements, and coordinate procurement activities. The business units themselves decide whether or not to participate in these activities (Boutellier/Locker 1998, p. 129).

To improve the coordination and cooperation and to make it possible to adopt joint measures for reengineering the processes in the SC, associations of the most important suppliers (supplier associations) have been established (cf. Rich/Hines 1997).

Depending on the procurement policy realized, distinct SC with different material, information, and financial flows result. Close cooperation in the SC can usually be realized more easily if central procurement policies are followed (Monczka et al. 1998, p. 74).

United Technologies Corporation (UTC) decided to integrate sourcing activities across its six major divisions and to make SCM a core competency for the corporation. The group had approached SC integration through cross-divisional purchasing councils. Although the councils were very effective in getting the various units to share information and knowledge, they were still operating within a highly decentralized procurement structure (meaning they could create corporate supply agreements but could not always deliver the volumes required to suppliers). To drive the transition, the firm has built aggressive cost savings goals into its business plans, thus providing strong incentives for plant and process managers to join the SCM initiatives. An extensive SCM training process was deployed to equip UTC personnel with skills specific to its long-term strategy. For the senior executives, much of the training focused on the human resource implications of the SCM strategy, such as the redeployment of personnel, reallocation of resources, and changes in skill sets required for various jobs (Porter 1998).

In applying traditional procurement policies, the quantities to be delivered are usually specified exactly in contracts. Intensive cooperation in SC results in outline agreements that provide the partners with midterm security about the volume of business, but also provides flexibility for fetching goods and services at short notice. To simplify the transport logistics, some outline agreements allow the individual delivery volumes to differ within defined limits from the ordered ones. In other cases, the logistics are simplified by a constant delivery quantity or by delivery in containers to avoid repacking. Contracts allowing for flexible volumes can dampen the transmission of order variability throughout the SC, thus potentially retarding the bullwhip effect (Tsay/Lovejoy 1999). Sometimes long-term contracts specify annual price reductions; in this case both partners have to improve the competitiveness of the SC.

Reorganization efforts often concentrate on indirect purchasing processes. Typical objectives of such reorganizations are (cf. Gebauer/Segev 2000):

- End users are empowered to select items and services. Allowing them to generate requisitions and purchase orders and to handle receiving and payment may save time and administrative overheads and reduce stocks and the associated inventory costs.
- Centrally provided multivendor stores help to guide end users towards a set of pre-contracted items. Cost savings result from discounts that are often reflected in prenegotiated arrangements. Large, decentralized companies expect high savings from renegotiation of long-term contracts with suppliers.
- Better documentation of indirect buying provides better data on spending patterns and process transparency. This information can be used for improving supplier management, for additional cost savings, and to enable fast reactions to changing environments.

If reengineering of purchasing processes leads towards decentralization, the decentralized activities are often coordinated by applying workflow management systems (Gebauer/Schad 1998).

2.3.3 Intra- and Inter-Company Organizational Changes

Strategies for basing the procurement on SCM concepts appear under such concepts as "Procurement Reengineering", "Supplier Alliance Teams" or "Strategic Supply Side Management". The Center for Advanced Purchasing Studies (CAPS) is engaged, among other topics, in changes of and forecasts about future procurement management [http://www.capsresearch.org/].

The close cooperation in an SC may foster "open-book policies" that enable a customer to view cost accounting data from its suppliers (Seal et al. 1999). An agreement on methods of cost allocation may be necessary to provide meaningful and comparable data. There have already been interventions from customers directed at getting a supplier to change its accounting procedures to provide more meaningful cost data (Carr/Ng 1995).

An intensified cooperation between the purchasing and sales departments should mean that customer requirements are better matched with the conditions in the procurement market. Furthermore, the concentration on the business activities counterbalances the technical orientation that often dominates, particularly, in SME. Sometimes, for example in retailing, the Efficient Consumer Response concept (cf. section 2.2.4) is used as a basis for combining sales and purchasing activities for each strategic business unit.

Karstadt AG, a leading German department store, defined new responsibilities for retailing and the manufacturing industry in its Quick Response Service. The quantities sold are reported directly to the suppliers via telecommunications. These suppliers use models to forecast the future consumption and so deliver specific quantities to the branches unless they veto the planned data. The products are packed in a form appropriate for the branches and are marked with their prices (Oertel/Abraham 1994, pp. 188).

Possible forms of cooperation in the procurement area are (cf. Baumgartner 1991, p. 74):

- Master agreements, letters of intent
- Guaranteed acceptance of minimum quantities or minimum volumes of sales
- Consigned goods located in the customer's warehouse but still the property of the supplier
- Quality control according to customer requirements on the supplier's site
- Provision of tools, equipment, and data
- Financing of investments
- Joint product development
- Selling products of suppliers under the company's name
- Storage on the supplier's site and JiT delivery
- Spare parts warehouse on the supplier's site
- Service provided by the supplier
- Influencing product policies of the suppliers
- Paperless ordering and communication.

The reorganization of SC often avoids redundant tasks and thus results in modified business processes. A case study describes the changes that Thermo King implemented when SCM triggered the migration to new material management principles and IT systems (Langan et al. 1996). Figure 2.6 shows a stepwise simplification of logistic chains achieved by the Zellweger Group, Switzerland, with its CATENA (Italian word for chain) project. If neither partner in the logistics chain wishes to assume the responsibility for these merged tasks, a third party (also for reasons of symmetry) could be made responsible for them; for example, a specialized service provider could perform the quality control using certified processes. Thus, the reorganization of SC does not always avoid intermediaries and new forms of intermediation sometimes arise.

Elida Fabergé is responsible for the personal hygiene business within the Unilever Group in Germany. On the basis of a study initiated in 1996, it began to orientate its business processes towards Efficient Consumer Response (cf. section 2.2.4). The realization of the logistics projects required the planned implementation of SAP R/3 to be brought forward by more than one year. The principal tasks of the "Supply Flow Management" department are the

- *definition of logistics concepts for suppliers using the knowledge gained in Elidas own sales department,*
- *optimization of procurement logistics and solution of material flow problems,*
- *quality improvement and coordination of supplier evaluation and development,*
- *optimization of order quantities, order cycles, and delivery rhythms, and reduction of material losses, e.g., through improved control of residuals.*

One of the achievements was improved coordination with the suppliers of packaging materials. Previously the choice of delivery times was left to the suppliers and carriers. This resulted in widely fluctuating utilization rates and temporary bottlenecks at the goods receiving and storage areas. A dialog with the ten most important suppliers of packaging materials led to a delivery calendar that avoided the earlier irregularities. There were next to no waiting times for these major suppliers. One day was reserved for smaller suppliers.

The suppliers' relatively limited interest in SCM is criticized. In future Elida Fabergé will orient its strategic supplier development and selection more closely on SCM aspects (Schulte/Hoppe 1999, pp. 73).

Vendor Managed Inventory (VMI) is a concept according to which the supplier assumes responsibility for inventory levels and for whether particular customers' goals (e.g., service levels, inventory turnover rates) are achieved. The main factors that could motivate suppliers to take over the warehousing function for selected customers are closer relationships, the associated competitive advantages, early recognition of changes in the markets, and the elimination of one decision echelon. The supplier receives early information about demand variations and can adjust its production planning and scheduling accordingly. Model analyses indicate that the improvement potential of VMI is remarkably higher than that of information exchange (Aviv/Federgruen 1998). However, an inventory reduction in the customer's warehouse often increases stock levels on the supplier's site.

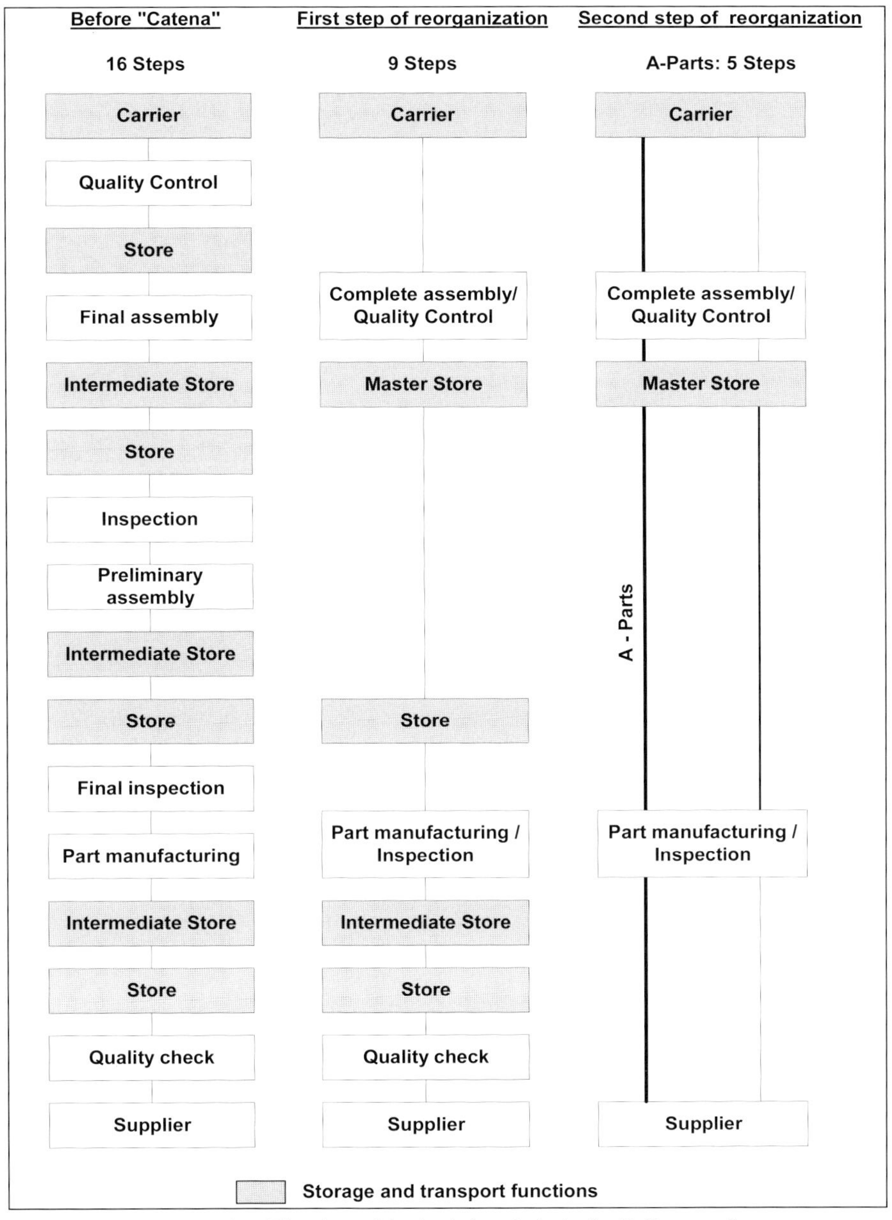

Figure 2.6: Simplification of the logistics chain in the Zellweger Group (cf. Schaumann 1991, p. 132)

Application of VMI requires that suppliers are familiar with the customers' inventory levels, up-to-date demand figures, and sales forecasts; they must also be able to supply small order quantities quickly and economically and operate IS that support VMI (cf. Tyndall et al. 1998, pp. 78). VMI is a type of the "Resident Supplier

Arrangement" in which members of the supplier's staff also act, for example, as contact persons for product development teams of the customers. The term "Co-Managed Inventory" emphasizes that responsibility may be shared between cooperating partners in the SC (Christopher 1998, p. 196).

In contrast to these viewpoints, "laws" of SCM have been formulated that consider the concept of the VMI rather critically (cf. Smith 1999):

- No company can manage a customer's inventory better than the customer can
- There is no free lunch. If a supplier manages a customer's inventory the cost of the product to some customers will go up
- The more closely inventory management is slanted toward the ultimate customer
  - the more stock keeping units will exist
  - the less accurate the inventory records will be
  - the less accurate the forecast will be
  - the shorter the replenishment time will be
- When suppliers are managing customer inventories, only simple systems work
- When a supplier manages a customer inventory, the customer may consider the inventory to be on consignment
- The difficulty of communication increases with rising number of suppliers each managing a segment of a customer's inventory
- Most suppliers can manage inventory for some of their customers; few, if any, are able to manage it for all of them.

The dm-drogerie markt, a major drugstore company in Germany, started an initial VMI project with Colgate in 1994; Colgate has been operating a VMI for dm since October 1996 (cf. the case study in section 5.1). This experience was used to draft guidelines that include a comprehensive process model consisting of four steps. In the meantime, approximately 20 percent of the revenues are based on VMI articles. The transition to VMI has not always led to substantial improvements in inventory and service levels. However, the manufacturer gains information on the availability of its products, can orient procurement and production processes more closely to the demand, and schedule its transport more tightly. This is in the interests of both partners. "Preferred suppliers" are defined to intensify the partnerships; the cooperation capability of vendors becomes an important criterion in selecting preferred suppliers (Rodens-Friedrich 1999, pp. 213; cf. also the Elida Fabergé example above).

The inventory, order, and sales forecast data needed for the VMI planning are normally transferred by EDI. In the SAP environment, they can be stored persistently in the APO. A VMI customer is created in the supplier network of APO as "location" and linked with the rest of the SC using a transport relationship defined in the Supply Chain Engineer. The planning process handles such a customer in the same way as a distribution center (cf. section 3.1.1).

If any of the VMI customers fails to provide sales forecasts, the supplier must calculate its own prevision for this client. This increases its planning effort and reduces the planning reliability.

2.3.4 IS for Procurement

CALS

The CALS concept has been used since the mid-1980s to systematize and standardize the computer support methods used to build logistical chains. This acronym was originally used by the American Department of Defense for "Computer Aided Logistics Support", later for "Computer Aided Acquisition and Logistics Support", and then for "Continuous Acquisition and Lifecycle Support"; recently it is used also in the sense of "Commerce at Light Speed", indicating a repositioning towards EC (cf. section 2.2.4). "The goal of the CALS Initiative is to enable integration of enterprises on a worldwide basis. The vision is for all or part of a single enterprise (e.g. an original equipment manufacturer and its suppliers, or a consortium of public and private groups and academia), to be able to work from a common digital database, in real time, on the design, development, manufacturing, distribution and servicing of products. The direct benefits would come through substantial reductions in product-to-market time and costs, along with significant enhancements in quality and performance" (Casey 1996).

Virtual Purchasing and E-Commerce

Virtual Purchasing means the use of the Internet as a platform for procurement activities, namely "buy-side electronic commerce".

General Electric (GE) opened an Internet access to its Trading Process Network (TPN) for its 25,000 suppliers in 1996. The suppliers can use the Internet to obtain information on parts tendered for procurement with specifications and drawings, and to submit their bids. The system makes more than two million drawings available in encrypted form. In this way, GE processes approximately 500 requests each day. The effort involved in preparing quotations has been dramatically reduced (Boutellier/Locker 1998, p. 158).

GE TPN Post is a secure Internet solution allowing buyers and sellers to conduct business electronically. It offers a robust solution combining software and services to help buyers locate new suppliers worldwide, streamline their purchasing processes, and dramatically shorten cycle times. The TPN Manager allows buyers to [http://www.tpn.com/]

- *identify qualified suppliers worldwide,*
- *distribute Requests for Quote and specifications to global suppliers,*
- *transmit electronic drawings to multiple suppliers simultaneously,*
- *hold multiple bidding rounds until a favorable price is reached, and*
- *receive and manage bids and seller communications efficiently and cost effectively [http://www.gegxs.com/gxs/products/product/getpnp/].*

In the following we consider the changes that result in electronically supported transaction processes, differentiating between the phases of initiation, agreement, and processing.

1. Initiation

The Internet has improved the transparency of many markets. Because easily described and communicated procurement criteria are very important for purchasing over the Internet, intensified price competition with negative consequences for product quality and innovation is feared. Some disadvantages of decentralized procurement can be reduced or even avoided by providing information about experiences gained with different suppliers over an intranet. The systematic collection of such information is an important task for a company's knowledge management.

SAP supports the procurement of add-on systems by short-listing on its web pages references that may be of interest for certain IS tasks. SAP certifies some of these solutions, providing a quality seal for technical compatibility. Customers may find additional information about the systems on the web pages of these vendors.

The purchasing department can use the intranet to provide guidelines for procurement activities. The University of Minnesota offers detailed help for decentralized purchasing designated as "virtual buyers" (NN 1999a). The Internet and CD-ROMs provide comprehensive information on potential suppliers that previously was often difficult to obtain in supplier reference works ("yellow pages"). In May 2001, the Thomas Register of American Manufacturers contained data about roughly 168,000 companies [http://www1.thomasregister.com/]. The information provider "Wer liefert was?" [http://www.wlwonline.de/] presents information about circa 370,000 companies from 13 European countries in printed form, on CD-ROM, and on the web. Prospective customers can use the Internet to formulate requests to specific companies, which are then forwarded by the intermediary.

At the end of 2000, about 1000 E-marketplaces had emerged or announced their future existence. A B2B marketplace is a virtual market where buyers, suppliers, and distributors find and exchange information, conduct trade, and cooperate via portals, trading exchanges, and collaboration tools. A new dimension is being introduced into SCM, with the formation of new partnerships causing competitive pressure while also creating a new level of collaboration between companies (Shams 2000).

The relevance of the B2B marketplaces (in this case for the chemical industry) is emphasized as follows (NN n.d.): The SC of the global chemical industry is characterized by significant inefficiencies. These inefficiencies are largely the result of manual transaction processing and inefficient use of information or resources preventing optimization of logistics across the industry. As a result, there are excessive resources, redundant inventory positions, and underutilized assets employed during the delivery of a product, with many firms providing similar services and little standardization. There is a significant opportunity to remove these inefficiencies from the industry via EC - it is estimated that in 2005 cost savings of $ 5 billion could be realized compared with the present. The marketplace was conceived and developed by 22 chemical industry leaders to eliminate the chaos in the SC and to make it easier for everyone involved to do business.

Commerce One describes the advantages of contracting with its marketplace as follows [http://www.commerceone.net/about/]:

- *Benefits to suppliers*
  - *Add a new sales channel with minimal start-up costs*
  - *Boost revenue potential by increasing the buying community*
  - *Eliminate the manual processes of managing operational resources*
- *Benefits to customers*
  - *Easily source suppliers, products, and services*
  - *Streamline the procurement process and reduce operating resource costs*
  - *Control buying by limiting procurement to contracted supplier.*

Services offered cover [http://www.commerceone.net/businessservices/]:

- *Collaboration in live and interactive web meetings, resulting in data, audio, and video communication via a web browser*
- *Catalog and content services help to define, develop, and cleanse catalog data for easy upload into XML-based catalog format and keep categorization, descriptions, and prices up to date*
- *EDI services with two major EDI Value Added Networks*
- *Data management and analysis*
- *Financial services (invoice, payment advice, and tax calculation)*
- *IT services*
- *Marketing and promotional services*
- *Product sourcing outside the Commerce One.net zones*
- *Auction services*
- *Shipping and logistics services*
- *SC Planning and Execution, defined as offering services such as key supplier adoption and integration of ERP and CRM systems throughout the SC.*

Some E-marketplaces have "ecosystems" that may help them grow: SAP claims that its ecosystem consists of more than 10 million users, 12,000 customers, 20,000 installations, 900 partners, and a customer base in each of more than 100 countries. It is assumed that only a few marketplaces per industry will survive and that SAP, with its experience in ERP systems, definitely has a very good chance of surviving the widely forecast B2B marketplace shakeout (Shams 2000).

Web-based product catalogs assume a central role in B2B relationships. The United Nations/Standard Product&Services Classification (UN/SPSC) developed by Dun & Bradstreet is an open standard that provides a framework for classifying goods and services. The system can be used to identify and classify products consistently for expenditure analysis, EC, card transactions, and supplier sourcing [http://www.dnb.com/prods_svcs/display/1,1318,DNB=132-FIR=1,00.html].

Compared with traditional, linear structures of information, the hypertext technology provides major advantages for establishing product catalogs. Powerful search engines are essential for retrieving information from catalogs. Potential customers could be interested in using the product catalogs to obtain data, to add comments, and to transfer selected data to their own IS.

The "Open Buying on the Internet" (OBI; http://www.openbuy.org/obi/about/) Consortium, founded in 1996, with a membership of circa 50 companies (as of 2001-05), developed several releases of OBI standards; release V2.1 was issued in May 2000. The specification enables the automation of high-volume, low-cost transactions between trading partners that account for a large percentage of most organizations' purchasing. The standard aims at reducing transaction costs and processing errors for both buyer and seller organizations as well as increasing efficiency and convenience of the overall purchasing process.

The open catalog interface of the SAP Business-to-Business Procurement module (cf. section 3.2) permits access to online catalogs.

Internet-supported purchasing leads to a conflict of interests between market participants, specifically as to whether product data catalogs should be stored and maintained by the suppliers

- on their own web sites (sell-side catalog; in this case the customers must search the web for unstructured information) or
- in accordance with the specific customer requirements in intranet structures of their customers (buy-side catalog). This procedure results in high data manipulation efforts of the suppliers. In this model the web permits bids to be obtained using electronic tendering.

One possible solution to this dilemma would be to transfer structuring and maintenance of the product databases to intermediaries. Each supplier and customer would then only have to establish interfaces to the providers that manage the product catalogs and would not need to implement an individual interface for each B2B relationship. Providers of so-called Operating Resource Management (ORM) systems, such as Commerce One, offer management of these data. Hybrid solutions can also be envisaged, with the master catalog containing all (also graphic) details remaining with the supplier and the customers storing only the most important product data. In this case, an update cycle appropriate to the business relationship must be agreed on (Bussiek/Stotz 1999, pp. 42).

SAP Markets, a subsidiary of SAP AG, and Commerce One have agreed on a close cooperation to deliver so-called next-generation marketplaces. In this partnership, Commerce One is responsible for organizing the infrastructure needed to build E-marketplaces, while SAP provides its expertise in business applications. Commerce One's cXBL (Common XML Business Library) is being adopted in the jointly developed products and will be extended to include SAP's Business Application Programming Interfaces (BAPIs; cf. Moser 1999).

In August 2000, SAP Markets and Commerce One announced the products MarketSet and EnterpriseBuyer. MarketSet is designed for customers who will set up their own branded E-marketplaces in order to manage communities of users and transactions. It claims to provide the following capabilities:

- *Integration with theoretically any buying, selling or enterprise application*
- *Intermarketplace transactions*
- *Auctions*
- *Multiple payment methods*
- *Content sourcing and services.*

EnterpriseBuyer provides capabilities for procurement of direct and indirect goods and also manages the SC cycle (Shams 2000). Future releases will link the trading systems to BOM systems and SC forecasts.

A competing EC consortium has been established by Ariba, i2 Technologies, and IBM in 2000. However, this collaboration was already revoked in May 2001 because some members extended into the core businesses of their "partners" (Evers/Oerlemans 2001).

Several goals may be relevant in the procurement of complex products. Emptoris [http://www.emptoris.com/] is offering multiple-criteria-based sourcing solutions that allow

- information sharing,
- buyer-defined purchasing policies,
- the application of business rules,
- customizable workflows,
- data tracking and analysis,
- what-if analyses, and
- the application of optimization techniques.

Agent-based comparisons support purchasing over the Internet. Agents are programs that move in intranets or parts of the Internet and perform services on behalf of a user. For example, they can be used to search the web for terms with which the same or similar parts are offered by different suppliers. Changes in the layout of web pages or special offers can cause problems with the automatic provision of data. For this reason, a standardized presentation of such information is desirable for the customers and their agents. Suppliers that block such standardization efforts reduce the transparency provided by the Internet. However, they also run the risk of not having their products sufficiently well publicized to prospective customers. To achieve interoperability between agent platforms from different suppliers, the Object Management Group (OMG) has proposed a standard for mobile agents (MASIF - Mobile Agent System Interoperability Facility; http://www.fokus.gmd.de/ research/cc/ecco/masif/).

The agent technique can also support the cooperation of independent procurement departments to form virtual purchasing cooperatives. The capabilities of some software robots exceed the price comparisons normally provided by "shopping bots" and can to some degree communicate with users in natural language (e.g., robots from Artificial Life [http://www.artificial-life.com/]). Research in the area of EC is concerned with the use of agents in negotiating processes; the "bazaar concept" is based on a sequential decision-making model that also considers learning processes (Zeng/Sycara 1997). A topology of multi-agent E-marketplaces is based on the following distinctions (Kurbel/Loutchko 2001):

- Type and number of negotiation partners
- Single- or multi-issue negotiation
- Crisp or fuzzy preferences.

Purchasing agents are offered, e.g., by Computershopper [http://www.zdnet.com/computershopper/]. Pocket BargainFinder, an agent developed by the Center for Strategic Technology Research of the former Andersen Consulting, is a handheld device that allows a customer to find an item in a physical retail store, scan in its barcode, and search for a lower price online. The device, a personal digital assistant with wireless access capability, combined with a portable barcode scanner, allows customers to physically inspect products while simultaneously looking online for lower prices (cf. Brody/Gottsman 1999).

An interesting coordination issue exists in multilevel SC. Let us assume that a buyer needs 100 components and approaches three suppliers to quote for them. A common 2nd-tier supplier is asked by each of the three to supply a quotation for a necessary subcomponent. However, this supplier can deliver only 100 units. If it is unaware that all three enquiries originated from a single original request and that only one batch of 100 subcomponents will be actually needed, the supplier will decline two of the requests and thus eliminate two of the potential suppliers. With multiple subcomponents it is possible that essential subcomponents will be locked up by different suppliers and that finally no supplier will be able to complete its bid. If, however, the buyer had been seeking 300 components and had split the total requirement among three suppliers, then there would be a demand for 300 subcomponents and the restricted availability of only 100 units would be a real problem. The MAGNET agent uses codewords to avoid such problems (Dasgupta et al. 1999).

XML is a control language for documents that contain structured information; cXML is an extension to XML to support EC. This language provides an XML-based infrastructure that can be used for secure handling of data exchange, update, delivery, and control processes for catalog content and transaction processes.

cXML supports "all supplier content management models, including buyer-managed, supplier-managed, content management services, electronic market-places, and Web-based sourcing organizations" (NN 1999c). The protocol permits suppliers to produce customer-specific catalog content and to differentiate them-selves from competitors.

Creditworthiness evaluations may be necessary prior to conclusion of contracts. Although such information was usually obtained (e.g., from Dun & Bradstreet) by fax, it is increasingly also being called up electronically. This external information may be passed to the vendor master data of the ERP system if its data model sup-ports this concept. Because these data cannot be corrected automatically within the company, they should be timestamped. An update in the database of the external information provider may cause a message to be sent to the information recipient, who then decides whether or not to update the locally stored information. It is also possible to take out subscriptions permitting changes to external data to be passed on automatically to an ERP system. This is one of many cases in which an tempo-ral database with timestamp attributes (cf. Snodgrass 1995) is desirable.

In 1999, SAP and Dun & Bradstreet (D&B) announced a strategic partnership for the integration of online business data in R/3. "D&B for SAP R/3" permits the user to augment its own information on business partners with data from D&B and so put its risk management on a more reliable basis. D&B for SAP R/3 pro-vides a customer and supplier solution in which the master data from R/3 and D&B's database are integrated at the application level. D&B also offers SAP cus-tomers a data rationalization service for aggregation, correction, and deletion of these data. With this strategy D&B extends the D-U-N-S number that identifies more than 50 million companies worldwide as standard for resource planning for maintaining customer relationships and for decision support.

2. *Agreement*

A further step in the direction of EC (cf. section 2.2.4) is done if electronic support in the agreement and ordering phases is available via the Internet. The legal valid-ity of electronic contracts is not universally accepted. The solution of security problems is of major importance for the success of EC (Buxmann/König 2000, pp. 41). In addition to traditional methods, digital signatures, actions certified by trust centers, and biometrically based techniques for user identification will become in-creasingly important in the near future. In routine B2B transactions the authentica-tion plays a less significant role than in B2C relationships, which are often only sporadic.

Private nets, VAN services, and the Internet can be used for the transmission of EDI messages. Cost comparisons differentiate between initial costs, fixed costs, and usage costs. The Internet has fixed cost advantages over VAN and advantages for both other cost types compared with the two alternatives. Such developments as Internet offerings from VAN providers, transmission guarantees from Internet service providers, and Virtual Private Networks are overcoming disadvantages of data transfers via the Internet (Alpar 1998, pp. 279; Faber 1998, pp. 229). In addition to electronic data transfer, agreements over the Internet will also be used in business relationships that have employed traditional forms of communication in the past.

The Business Application Software Developers Association (BASDA) announced an initiative in June 1999 to introduce an international standard business document interface for electronic business [http://www.basda.org/]. ebXML is a set of specifications that together enable a modular, yet complete electronic business framework. The ebXML architecture provides

1. a way to define business processes and their associated messages and content,
2. a way to register and discover business process sequences with related message exchanges,
3. a way to define company profiles,
4. a way to define trading partner agreements, and
5. a uniform message transport layer

[http://www.ebxml.org/white_papers/whitepaper.htm].

In May 2001, ebXML was approved by UN/CEFACT, the United Nations body whose mandate covers worldwide policy and technical development in the area of trade facilitation and electronic business, and by the Organization for the Advancement of Structured Information Standards (OASIS), a consortium that advances electronic business by promoting open, collaborative development of interoperability specifications. This means, in particular, that SME will be able to replace relatively expensive EDI and VAN systems with web applications or to introduce such solutions for the first-time.

3. Processing

Procurement actions initiate flows of services and goods, information, and finance. Some of them are supplied in digitalized form and may thus be transported via the Internet; these include, in particular, providing text, graphics, and audio and video sequences. Irrespective of whether or not the service or good is provided digitally over the Internet, accompanying tasks arise in the processing phase that often can be supported electronically.

When electronic order processing is used, it should also be possible to make payments over networks (electronic cash; cf. Buxmann/König 2000, pp. 36). In B2B relationships, because other mechanisms are available and there is often a relationship of trust between long-term business partners, payment data are normally not transferred via the Internet.

The SAP Business-to-Business Procurement component allows electronic processing of all procurement tasks, starting with an inquiry and ending with payment of the subsequent invoice over the Internet. Section 3.2 describes this component in more detail.

The SAP Online Store provides a facility for manufacturers and retailers in which they can offer their products in an electronic catalog. The catalog permits a multimedia presentation of the products. These can be assigned to categories and displayed in hierarchical structures. Selection fields are provided for variants (e.g., by size or color). This representation simplifies the maintenance of the catalog pages. Search engines available in the Online Store permit specific retrievals, e.g., by keywords or product names (Figure 2.7).

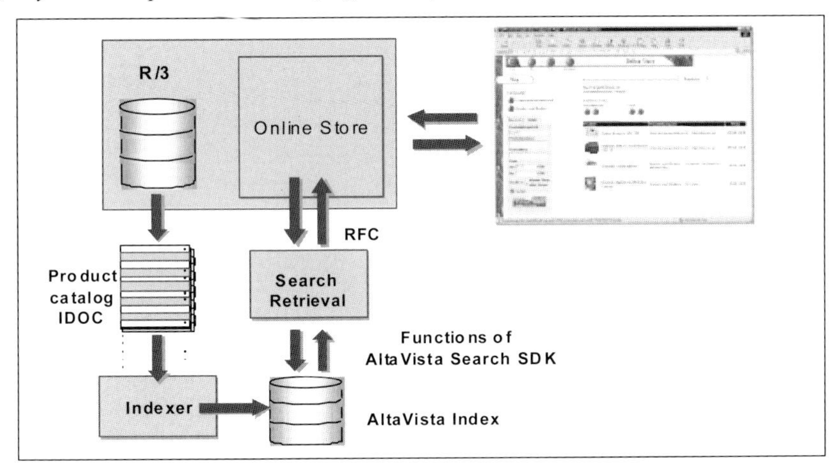

Figure 2.7: Use of the AltaVista search engine in the SAP Online Store

Online Stores may be created (possibly in different versions) for both B2B and B2C relationships. The store allows the customer to query the current prices and the availability of the goods simultaneously. The supplier's ERP system transfers the required data to the Online Store. Offers and order confirmations can be handled via E-mail. When orders are entered they initiate R/3 transactions for further order processing.

Figure 2.8 shows how SAP may connect various tools for Business-to-Business Procurement in a supply network. The manufacturer obtains information from the SAP Online Store or from other electronic product catalogs about alternative materials. Figure 2.8 shows examples: Commerce One (C1) and Intershop systems providing product catalogs, and a Harbinger Corporation system. The last company provides industry-oriented basic data and catalog structures into which the suppliers can place data relating to their products.

In addition to the procurement of physical assets, many other procurement tasks arise in companies, e.g., for capital and services, which are often processed without using IS. ORM systems should support these procurement tasks. Well-known systems are ORMS from Ariba, which, for example, links Nestlé USA with SAP R/3, and the Commerce Chain Solution from Commerce One.

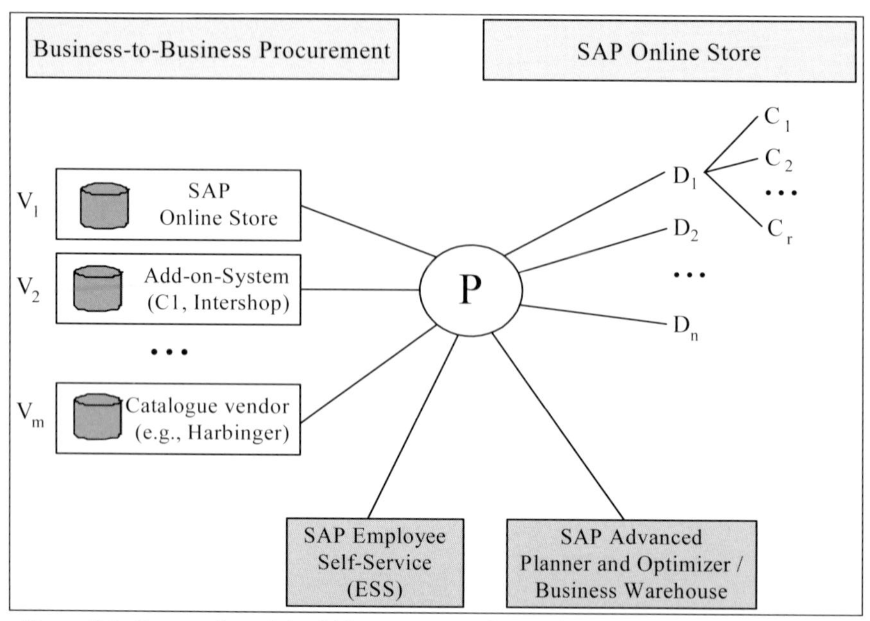

Figure 2.8: Cooperation of the SAP components for Business-to-Business Procurement with suppliers, manufacturers, distributors, and customers
Legend: D .. distributor; P .. producer; C .. customer; V .. vendor

Many E-markets concentrate on self-service buying of rather trivial products, ranging from simple office products to MRO items. The procurement of these indirect, non-production-related items is usually organized quite differently from the sourcing of direct, production-related materials (Gebauer/Segev 2000). For procurement of MRO products, often no real-time integration of systems is implemented. Exchange of data with these suppliers is usually asynchronous and message based. Desktop and other web-based purchasing systems must be linked with internal ERP systems (Dolmetsch et al. 2000, pp. 215). Real-time access to data in an ERP system requires disclosure of the application program interfaces (API). All major vendors of desktop purchasing systems claim to be able to realize a real-time link to the leading ERP systems.

SAP is pursuing the concept of business application program interfaces that have been semantically defined on the basis of the reference model. These BAPIs are used by the vendors of desktop purchasing systems for extracting data from SAP R/3 and writing other data back.

Virtual auctions are a highly developed form of virtual procurement. The first auctions on the Internet took place in 1995, and since then auctions have evolved to the point where they have become an important element of electronic markets (Schmidt et al. 1998). With Internet auctions there are no costs for offices, personnel, and travel; there is also often less time pressure on the decision-making process than at conventional auctions.

Companies looking for a certain product may either perform a full-text retrieval or a specific search via product and/or manufacturer names. To identify interesting offerings, the auction house can provide a similarity-oriented search ("find similar products"). Product information is presented directly on the web site or made available by links. Many auction forms, such as English, Dutch, First or Second Prices Sealed Bid Auctions are used in the agreement phase (Schmidt et al. 1998). In most auctions the bids must be changed within predefined intervals (ticks). The interested party finds the best current bids ("winning bids") on the web sites of the auction house. E-mails may be sent to interested parties who have been outbidden. The end of an auction can be determined in several ways: when a certain time point is reached, or by a decision of the auctioneer or of the supplier.

Currently SAP systems do not support the processing of or participation in virtual auctions. In the future, electronic tendering procedures will permit companies to publish their tenders on the Internet. Suppliers can respond to these tenders with appropriate offers. It may be possible for selected suppliers only (restricted tendering) or for any company (public tendering) to place orders.

Sometimes discrepancies arise within the many documents that accompany a shipment transaction. eTime Capital [http://www.etimecapital.com/] tries to improve the "financial SC" by a web portal that makes documents viewable by all participants involved. This allows customers to discover data discrepancies instantly and thereby speed up the flow of documents and cash.

An analysis of the customer's business processes identifies the full range of documents a sale may generate, together with which party creates it, and when. This information is passed on to two applications: A rules engine describes the range of errors and mismatches a customer has decided it will accept, e.g., in purchase orders, sales orders, change orders, and invoices. The software also defines the circumstances in which those deviations must be flagged and alerts sent to the party concerned. The other program is designed to extract relevant records from ERP systems. It tags the data in XML and passes it back to eTime's server, which uses the data to create its own version of each extracted document. The BusinessNow server can alert managers of problems by E-mail as well as highlighting disputed documents on view at the portal [NN 2000].

The implementation of an E-procurement solution that is reasonably integrated with backoffice systems and the relevant systems of the SC partner is a major project, for which thus far little experience is available. In the early 1990s the deployment of ERP systems was very challenging, and success and "war" stories appeared in the trade journals. Ten years on, a similar situation holds for E-business applications (cf. Gilbert 2000). However, in the same way as ERP systems have become routinely managed components of an IS architecture today, E-business applications will also be regarded as a matter of course in the near future.

In certain way, SCM and EC are at odds with each other: SCM tends to tighten business relationships between selected companies, whereas EC makes the offers available on the market more transparent and may thus lead to looser, more easily changeable relationships. SCM tends to reduce the number of suppliers, and EC may contribute to extending it. In fact, the supposed contradiction may be resolved if different types of products and services are considered. EC seems to be concentrated on purchasing non-strategic, less important, indirect goods, whereas SCM focuses on A-parts.

2.4 Production

The production department, and therefore also production logistics, adopt a key role in order management. Important goals of the production department are short lead-times, low inventory, on-time delivery, and high utilization rates of resources. Frequently the first three goals are in conflict with the fourth. The metrics mentioned above are relatively easy to observe and often used as substitutes for enterprise goals that are more difficult to quantify, e.g., the return on investment or customer satisfaction.

The former Digital Equipment Corporation has been working on the development of a global SC model at the strategic level since 1992. A mixed-integer programming model was used to prepare an integrated, global plan of the production and distribution structures that had to be realized within 18 months. A top manager from Digital explained in 1994: "Prior to 1991, we were making decisions out of several different structures, several different organizations ... the unfortunate thing is, they never came together into one decision. We had a large confusion factor." The restructuring of the SC resulted in cost savings in excess of $ 100 million (Arntzen et al. 1995).

Production is organized in various ways; depending on the organizational form selected, different numbers and types of interfaces occur in the material and information flows. Consequently, the sections below discuss different organizational forms and the resulting requirements for ERP systems.

2.4.1 Organizational Structures in Manufacturing

One effect of the interfaces between the individual areas of a shop floor is that the work objects are repeatedly placed in queues and the sum of the processing times consequently represents only a small percentage of the total lead-time. Flexible manufacturing systems are capable of performing operations that are traditionally assigned to different machines. Even if a certain type of production cannot be organized as flow shop, attempts should be made to shift job shop manufacturing in the direction of object-oriented flow shop production and, thus, towards SC.

Industrial production was originally organized primarily into supervisor areas and its management was thus decentralized. In this organizational form, those persons who are most familiar with the tasks and the current status of a work center also make the associated operative decisions. However, there is no information available about the production system as a whole, which, in the best case, leads to local optima; there may also be conflicts between company goals and division- or individual-related objectives.

When the early MRP (Material Requirements Planning; Manufacturing Resource Planning) systems became available on mainframe computers, this resulted in a more centralized, computer-supported type of decision making, although the mandatory character of the centrally determined plans was handled quite differently. The response of these systems was so slow that their results could not be updated daily, let alone in real time. Thus, locally available information, such as unplanned personnel absences, machine stoppages, or the need to correct faulty parts, was only available on the shop floor and not immediately available for central production planning and scheduling. In such a centrally organized production control, either staff follow the computer-generated instructions despite knowing better or they intentionally "work around the system".

Irrespective of computer response times, continually changing plans are not well accepted by employees, because they do not agree with "rescheduling the schedule just rescheduled"; therefore, ongoing changes are scarcely realistic. Plans must remain valid for specific time periods so that they can be realized; intervals ("frozen zones") must be defined during which modification is allowed only in the case of particularly significant events. Bureaucratic rules on quasi-unchangeable periods may be specified, or a price-controlled mechanism introducing penalty fees chargeable for changes made to schedules after they have been made known.

Other concepts see the focus of MRP support in mid-term planning and intentionally give decentralized units some freedom in short-term sequencing and scheduling. However, whatever refinements are incorporated, MRP systems cannot take account of all the knowledge available in the heads of the workers: For example, setup times are usually better known to the workers who perform the tasks involved than to the work schedulers. Setup time matrices that specify the required setup times depending on the most recently performed work step can hardly be determined exactly for widely used resources. This means that SCM must structure the knowledge available at various levels in the participating organizations in an appropriate manner, save it, and make it available on demand (Knowledge Management).

Decentralization concepts are aimed at regrouping the enterprise, on the basis of integrated, customer-oriented processes, into small, easily manageable units. These have decentralized decision-making competencies and responsibilities for the operating results achieved. Because the material flow and inventory levels are subject to physical control, the amount of accompanying paper documentation can be reduced. A modern form of decentralized manufacturing needs well-defined and well-implemented interfaces between the IS of the units.

Group work of several staff members regularly working closely together is intended to lead to higher motivation and increased flexibility. In the last few years, the reasons for considering this organizational form have changed from humanization of work to potential economic advantages. Appropriate mid-term planning has to ensure that the tasks allocated to a certain group can be realized in the time available. The sequencing and scheduling decisions of the group use these preliminary fixings together with immediate observations of its manufacturing area. Depending on the complexity of the processes to be scheduled, the decisions can be made without IS support or assisted by control stations (so-called "leitstands"; cf. Adelsberger/Kanet 1991; Pinedo 1995, pp. 304). Because group work means that the events, conditions, and actions of only locally relevant processes are no longer available in the central IS, the potentials for event-oriented data processing and for the application of business rules are reduced. The specifications of IS that support group working must define which data need to be available only in a particular group and which are also relevant to the work of other groups, functional areas or external members of the SC.

On the basis of a study initiated in 1996, Elida Fabergé began to organize its business processes according to the principles of Efficient Consumer Response (ECR, cf. section 2.2.4). As part of this strategy, partially autonomous groups were introduced in the production areas of its German plant in Buxtehude. Two of five hierarchy levels were eliminated and a change made from a functionally and hierarchically structured organization to a lean, process- and learning-oriented organization. In the production areas

- *the staff has been trained to increase their contributions to customer benefits,*
- *multiskilling has been introduced to enable staff members to be assigned more flexibly,*
- *the working times are determined by the group members,*
- *the sequences for execution of the operations are also specified by the group members, and*
- *a program to reduce the setup times has been initiated (cf. Schulte/Hoppe 1999).*

In SAP R/3, some procedures provided by the PP (Production Planning) module can be used to support work groups. Several workplaces may be combined to form a group. The completion messages required for the plant data collection take place either after the occurrence of certain events ("milestone confirmations") or after certain time intervals ("progress confirmations"). Should unexpected deviations occur, the group has the authority to initiate corrective work activities or to trigger reprocessing orders, to start workflows for information purposes, and to send the appropriate messages.

One of the main goals in organizing an SC is the reduction of complexity (cf. section 1.2.3). Manufacturing segmentation (Wildemann 1994, pp. 225), which is closely related with the group work concept, establishes "factories in factories" which are less complex and have greater autonomy, market closeness, and flexibility, leading to simpler information, planning, and control tasks. A segment includes several echelons of the internal logistical chain, or even all of them. Preconditions for successful segmentation are efficient information interfaces between the segments. To simplify the coordination, organizational units performing closely related tasks should be located near each other whenever possible. The implementation and use of shared databases and data warehouses should make information originating locally available in real time if it is relevant in other segments. Intranets can support the dissemination of this information, which may be organized according to either the push or the pull principle.

To coordinate the processes between manufacturing segments, "capacity exchanges" are recommended, in particular for solving conflicting demands for bottleneck resources. The supply corresponds to the capacity available, and the demand depends on the orders to be scheduled in a certain time period.

A highly segmented plant becomes a "fractal factory". A fractal should provide all workers with the capability to think and act to the benefit of the company, and should promote self-organization and self-optimization in small, fast-acting control cycles (Warnecke 1993). The role and importance of IS within and between the fractals has not yet been studied in detail.

2.4.2 MRP Systems and Beyond

MRP systems and the logistics modules from ERP systems provide interfaces, for example, to database management systems, to CNC, human resource, and accounting systems. The CIM concept, which is closely related to some parts of the SCM approach, has the goal of achieving an intensive connection between technical and business data processing. The integration of MRP solutions with IS of other functional areas results in cross-functional solutions that support business processes and intra-company logistics. This focus is sometimes emphasized by the term "logistics-oriented MRP system".

R/3 offers the PP and Project Management System (PS) modules for different types of production. The R/3 kernel is extended with industry-specific solutions called "Industry Solutions". There are also third-party add-ons to R/3, e.g., for the chemical industry, which use the Production Optimization Interface (POI). They check many conditions prior to the dispatching of each operation, and support, for example, decisions between additional setup times or longer throughput times.

If companies operate production facilities at several locations, the IS can reflect this distribution or be (at least partially) centralized. Distributed IS contain several, possibly heterogeneous components, but appear as a single local application to the user. In a distributed system, the users should be able to invoke all functions and data without needing to know where (on which network nodes) they are stored. In this regard, the conception of the currently available ERP systems is central, despite the physical distribution of application servers and clients. This applies even if databases are stored redundantly on multiple servers. The coordination of decentral MRP systems causes major conceptional problems (Corsten/Gössinger 1998). A possible coordination strategy that became available only recently is the decentralized use of ERP systems coordinated by an SCM system (cf. section 3.4.1).

Despite the wide use of MRP systems, particularly in large enterprises, they have several weaknesses (cf. Zäpfel/Missbauer 1993, pp. 299; Drexl et al. 1994; Fernández-Rañada et al. 1999, pp. 3, pp. 147; Norris et al. 2000, p. 90; Voß/Woodruff 2000):

- Lack of options for influencing the system behavior with respect to strategic and operative goals
- Internal focus, thus addressing only one node of the SC
- Limited decision support capabilities
- Scheduling decisions based on a priori specified lead-times, making insufficient use of information about the current situation on the shop floor
- Accumulated risk aversion of different planners, resulting, for example, from uncoordinated increments of lead-times that are expressed as decimal numbers with respect to the length of the time buckets
- Insufficient consideration of changing business environments, such as higher importance of customer-specified variants (mass customization, cf. section 2.2.2) and increased flexibility of production facilities
- Insufficient consideration of potential bottlenecks
- No support for making decisions on alternative means of reducing throughput times
- Inadequate plant data collection, resulting in poor data quality.

The weaknesses of MRP systems also make themselves felt in the logistics modules of ERP systems. This leads to the criticism that the business functionality of ERP systems has remained largely unchanged, with the main improvements being made primarily to the user interfaces. Some of the weaknesses mentioned above are eliminated by SCM systems.

The main differences between ERP and SCM systems are summarized in Table 2.2 (cf. Zheng et al. 2000, pp. 89).

Master Data for Production Planning and Control

MRP systems usually concentrate on mechanical manufacturing, assume the existence of master data in the form of bill of materials, routings, personnel and machine data, and are based on sequential planning concepts.

For mechanical manufacturing the direct quantity relationships between parts are usually described by single-level BOM. The structure of a particular product can be illustrated in so-called Gozinto graphs showing these BOM data at their edges. All other types of bills of materials (e.g., the multi-level BOM) can be derived from the single-level BOM, and the total requirement matrix may be derived from the assembly parts matrix (Vazsonyi 1954). These representations cannot always be used, e.g., if by-products flow back to earlier production stages.

This is why SAP R/3 was extended by special functionalities, e.g., for the process industries, such as SAP Chemicals and SAP Pharmaceuticals. Many firms have replaced specialized software packages for the process industry, such as PRISM, CIMPRO or Process One, with ERP systems that have been extended and adjusted specifically to meet the needs of this industry.The development department initially specifies the BOM. However, engineers often have views about relevant parts that differ from those needed, for example, in the production department or for after-sales services; thus, often department-specific variants of a BOM result.

Table 2.2: Differences between ERP and SCM systems

	ERP	SCM
Focus	Integrating and improving internal business processes, material flow, and information flow.	Integrating and improving internal business processes, material flow, information flow, and cash flow as well as the interaction of the organization with its business partners.
Major business areas supported	Financials Controlling Manufacturing Human Resources	Manufacturing Logistics Supply Chain Planning
Customer relationship	Reacting to customer demands, but no involvement of external parties in process improvement.	Involvement of external parties in process improvement; anticipation of customer needs and demands.
Scope	Coordinating and integrating all activities within a single organization.	Coordinating and integrating also all interorganizational activities.
Planning and execution support	Focused on internal planning; support of execution by decentralized and/or add-on systems.	Capable of controlling both planning and execution, both inter- and intraorganizational.
Focus of planning	Infinite planning without considering the limited availability of key resources required for executing the plans.	Finite planning providing feasible and reasonable plans based on the limited availability of key resources required for executing the plans.

Various views on a common BOM can be defined in R/3. It is possible to customize each application to determine which type of BOM is to be accessed at run time; this makes it possible, for example, to specify that cost accounting uses the costing BOM whereas advanced CAD techniques, such as solid modeling, have to access the design BOM.

Routings describe the operations needed to manufacture each in-house production part defined in the BOM. For each operation its number, the tasks to be performed, the dedicated resource group, its setup time, and the processing time per unit have to be defined. These data constitute not only the basis for multiperiod production planning, control, and execution, but also for accounting and certain types of salary systems.

A major problem with providing master data is that a large number of manufacturing options may exist, resulting in many BOM variants and/or routings (Schönsleben 2000, pp. 261). Often the large number of possible combinations cannot be modeled adequately in the master data. At least to some extent, decentralized control may compensate for the data that are missing in many MRP or ERP systems by using the specialized knowledge of production workers.

Production Planning

The independent demand is known once the sales program, internal orders, and planned changes of inventory levels have been specified. A BOM explosion determines the total demand for all parts to be produced or bought. This procedure uses the lead-times of the parts specified in the master data to approximate the time requirements for production and procurement processes. Demands occurring in different periods can be bundled into one production lot if, to keep the sum of setup times low, not all parts are produced in each time interval. The production quantities are temporarily assigned to certain time buckets in lead-time scheduling and to certain resource groups in capacity requirements planning; these planning steps yield a load analysis for each resource. Corrections may be necessary in the subsequent capacity-leveling phase. These can be performed either automatically or in a dialog with the planner and may, for example, change the available capacity of certain bottlenecks or the temporal assignments of production orders to resource groups. The goal is to determine a feasible production plan in which the available resources are not exceeded, except possibly for predefined small percentages.

Because the BOM explosion in MRP systems usually ignores the resource assignment, its results can cause execution problems. The master plan for the final products may need to be changed several times by trial and error before acceptable capacity loads are obtained. Such iterations are the price that has to be paid for sequential planning which, for reasons of simplicity, ignores potential bottlenecks in the first step.

For many decades, economic theory has discussed methods of bottleneck-oriented planning. Eugen Schmalenbach, the founder of German management theory in the early years of the twentieth century, proposed opportunity cost (shadow prices) as a mechanism for allocating bottleneck resources. Several methods developed in decision theory and Operations Research identify and assess bottlenecks. Many publications (e.g., Drexl et al. 1994) criticize the inadequate consideration of bottlenecks in almost all MRP and ERP systems.

There are several reasons for the reluctance to make practical use of bottleneck-oriented planning methods:

- Capacity limits must be defined as constraints. Because of imprecise data and the inherent flexibility of the production system, these constraints often do not exist in a very precise form. Small changes to inaccurate data can have significant effects on the optimum solution and result in major changes of the associated resource valuations (shadow prices). For this reason, sensitivity analyses and scenario techniques have to be applied.
- Changing business conditions may lead to large differences, also in bottlenecks. Fluctuating bottlenecks cannot be forecast adequately and make scheduling decisions cumbersome.
- Human actions are planned and executed at various hierarchical levels; many a priori uncoordinated adaptation measures may be implemented to avoid problem situations and bottlenecks. This flexibility can scarcely be adequately represented in planning models.
- The missing link between bottleneck-oriented optimization models and operative ERP systems is one reason why optimization has not been more widely used in the past. For many practitioners, these interfaces and the methods of generating and maintaining optimization models are more relevant than improvements of optimization algorithms.

The "Theory of Constraints" (TOC) has recently been subject of a great deal of interest. This method is concerned with finding economically suitable solutions in the presence of constraints and bottlenecks; it is based on the Optimized Production Technology (OPT) approach (Goldratt 1990; Dettmer 1997; McMullen 1998; Cox/Spencer 1998). APICS (The American Production and Inventory Control Society, now named Educational Society for Resource Management) organized a Special Interest Group on "Constraint Management" and has been holding annual symposia and workshops on the TOC since 1995. Several proposals developed in this context have been implemented in APS systems (cf. Tayur et al. 1999; Stadtler/Kilger 2000).

DaimlerChrysler started a constraint management pilot project at its Mercedes plant in Sindelfingen in spring 2001. Its goals are higher efficiency in planning and more sophisticated reactions to deviations from plans, fostered by simulation studies (Heitmann 2001).

The mySAP SCM system applies advanced, bottleneck-oriented methods of business planning: APO provides optimization algorithms and heuristics with many more features than the rather simple procedures available in MRP and ERP systems. The APO methods can be used to support the whole SC, but may be also applied for detailed production planning and scheduling of a single plant. Several software vendors have implemented solutions based on the TOC; SAP uses optimization engines developed by ILOG [http://www.ilog.com/] in its APO system. The methods now available in APO and other APS systems can overcome some of the deficiencies of conventional MRP and ERP systems (cf. section 3.1.3).

An APS system is described as follows: "Uses constraint models that treat both materials and capacity to set production priorities over time horizons of several months to several years. Advanced planning uses a dedicated server and in-memory processing, combined with special algorithms, to generate production plans cognizant of material, capacity, and other constraints as they are at that moment. Processing speed allows for flexibility in planning and lets users run simulations that base delivery promises on actual production conditions. ... The initial impact of advanced planning has been within enterprises, but advanced planning systems can be expanded to model an entire supply chain, including vendors, distributors, and points of demand. As vendors assemble product suites, they face challenges of their own. For one, advanced planning is not a single technology; different solver techniques are used depending on the kind of problem solved. And the realities of supply chains are so complex that even today's most powerful computers cannot model them without some level of aggregation. Finally, the need to integrate disparate computing systems has brought data structure problems to the surface that must be addressed" (MSI 2001).

Main features of an APS system are (Alvord 1999):

- Provides extensive modeling capabilities
- Handles complex, multi-level BOM
- Supports hard- and soft-pegging (cf. section 3.1.3)
- Supports hard and soft constraints
- Supports complex routings
- Provides plant-level ATP and capable-to-promise (cf. section 3.1.4)
- Provides what-if analysis of multiple production scenarios
- Identifies bottlenecks for critical resource management
- Ensures a smart release of jobs to reduce waiting times and enlarge bottleneck utilization
- Is bounded with industry-specific templates to ensure rapid implementation and deployment
- Contains interfaces to ERP systems for up-to-date material availability.

Interfaces between APS systems and other planning, scheduling, and execution systems are shown in Figure 2.9.

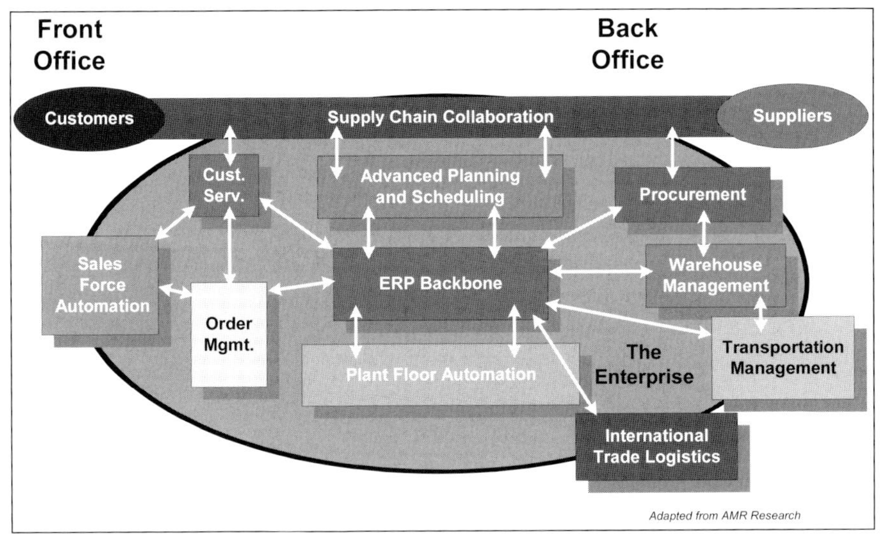

Figure 2.9: IS architecture showing interfaces between different planning, scheduling, and execution systems (cf. Bjorksten et al. 1999)

In addition to traditional optimization techniques, such as linear programming and mixed-integer programming, many heuristics have been implemented in APS systems. Comparing heuristics is a rather tedious task, and their quality is therefore difficult to judge. Different parameter configurations usually influence their relative performance. Even if serious experimental comparisons are provided, a trade-off has to be made between the mean solution quality, robustness (in terms of standard deviations of the differences between optimal and heuristically determined solution values for different problem classes), and the solution effort. In addition, the soundness of generalizing the results has to be critically examined.

Some of the solution approaches of ASP systems are vague combinations of already existing and new procedures; for example, methods proposed for solving the "constraint satisfaction problem" are classified as (Barták 1999):

- Systematic Search
  - Generate and test
  - Backtracking
- Consistency Techniques
  - Node consistency
  - Arc consistency
  - Path consistency
  - Restricted path consistency

- Constraint propagation
  - Look back
    - Backjumping
    - Backchecking
    - Backmarking
  - Look ahead
    - Forward checking
    - Partial look ahead
- Stochastic and heuristic algorithms
  - Hill-climbing
  - Min-conflicts
  - Random walk
  - Tabu search
  - Connectionist approach.

However, if determination of the optimal solution to a constraint satisfaction problem is the goal, the well-known Branch&Bound method is recommended.

"Constraint Programming" is a method for solving combinatorial problems; it has its roots in the field of programming languages and not in the area of Operations Research. Its goal is to reduce the development effort of combinatorial optimization algorithms. A constraint component provides the basic operations of the architecture and consists of a system reasoning about the fundamental properties of constraint systems (e.g., satisfiability). Operating around the resulting "constraint store" is a programming-language component that determines how to combine the basic operations, often in a non-deterministic way. In addition to such techniques as linear programming, interval reasoning, and Boolean unification, recent versions include algorithms such as edge-finder and its generalization for scheduling applications.

One of the disadvantages of constraint programming is the lack of support provided for modeling. The developers of the Optimization Programming Language OPL attempted to combine the strengths of algebraic modeling languages (e.g., AMPL [http://www.ampl.com/]) and of constraint programming (Van Hentenryck 1999, pp. 1). Comparisons show the close connections between constraint programming and integer programming methods (Bockmayr/Kasper 1998; Williams/Wilson 1998).

The developments described above raise the question of whether the acceptance of well-known optimization methods will be better, now they have been embedded in SCM systems, than it was with the earlier stand-alone optimization software. The following considerations are relevant in attempts to answer this question:

1. Compared with traditional intra-company planning, the organizational and mathematical complexity of planning increases when the entire SC is considered.

2. Nowadays, optimization models can be specified using a graphical user interface; the planner is partially freed from algebraic model formulation and does not need detailed knowledge of matrix generators, planning languages, and solution methods. Deviations from specified constraint values can be allowed by specifying penalty costs ("soft constraints"; cf. section 5.1.4). These may, e.g., be used to model additional, more expensive external capacities (so-called "premium capacities").
3. Many users try to avoid individually developed interfaces between their operative systems and supposedly complex optimization software. The conceptual and physical connections between ERP and APS systems greatly simplify access to operative data and the associated data aggregates. The vendors of ERP systems thus provide valuable integration services. These are expected to stimulate the acceptance of optimization methods and models.
4. Because of the dramatic improvements in hardware efficiency, large data volumes can now be kept in the working memory (cf. liveCache concept of SAP APO and section 3.1.3). This technology reduces the accesses to secondary storage media and is one of the main reasons why the application of complex solution methods has become far less time-consuming over the past few years. However, additional integrity problems may arise in case of system failures.
5. Managers often preferred a package with a graphical user interface based on a heuristic to one providing exact optima but missing a pleasing user interface (Geoffrion/Powers 1995). The widely used graphical, and, more recently, web-based interfaces permit visualization of the optimization processes and of the results obtained. This feature supports acceptance of the optimization results and of the delivering systems at different management levels.

Theoretically, the quality of the solution methods implemented in an APS system should be a major issue in decisions on deploying such a system: "Don't invest in an APS system until you really understand the nature and power of its optimisation engine" (Steele 1999). Unfortunately, the methods used for APS systems are not always described in detail, exotic names are introduced, and scientifically well-defined notions are abused (Knolmayer 2001b). "One of the most confusing aspects of the optimization market is the variety of solver methods marketed with esoteric names. By and large, vendors are diligent in using appropriate methods within their solvers - whether the methods are proprietary, purchased from another vendor, or based on known methods. This makes solvers a secondary consideration in choosing among various optimization applications" (Lapide 1998). On the other hand there are empirical results indicating that the methods implemented have become the main criterion in evaluating APS systems: "In contrast with the 1993 result, in 1999 the practitioners ranked optimality of the solution as the most important characteristic of the software" (Goetschalckx 2000, p. 84).

From the SCM perspective, production planning is considerably simplified if an intensive communication with the customers permits the demands to be better forecast and if the information exchange with the suppliers allows them to estimate their stock levels and delivery times more realistically. This cooperation can reduce the bullwhip effect described in section 1.2.1.

Similar consequences to those described by the bullwhip effect also occur in multi-level production systems for inventory policies based on reorder points (ROP) (cf. Figure 2.10). Uncoordinated lot sizing over several production stages often results in large lot sizes at low manufacturing levels. The average inventory is significantly more than half the lot size or half the order quantity; for this reason, coordination of dependent demand and lot sizes by MRP systems should be preferred to the use of ROP (Orlicky 1975, p. 26).

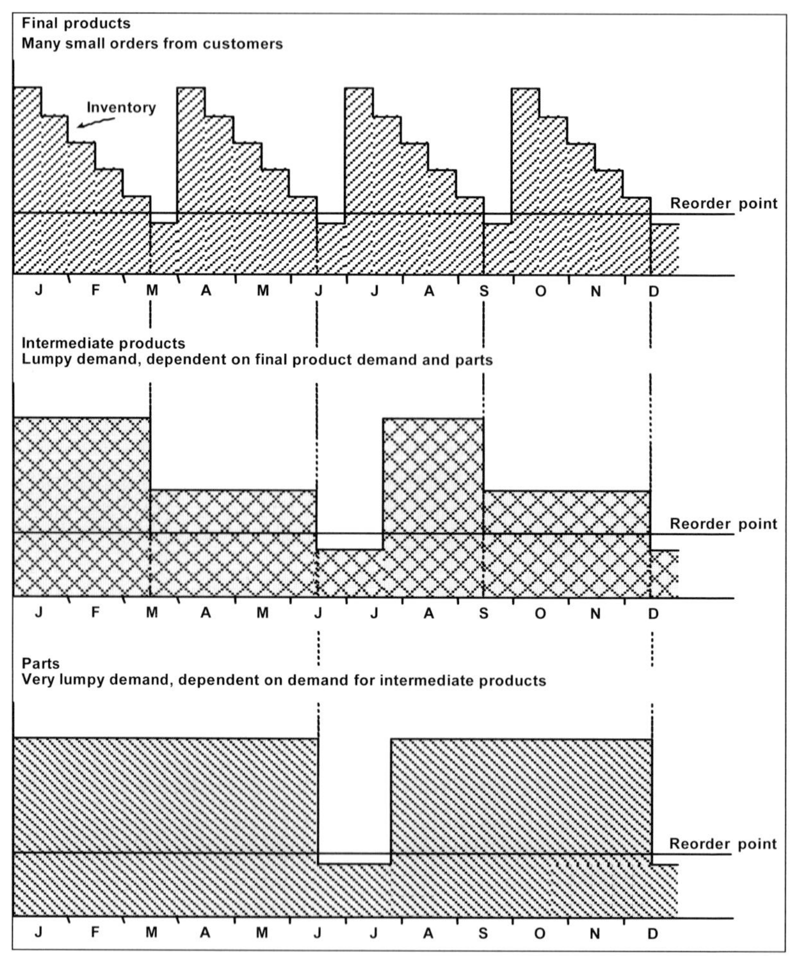

Figure 2.10: Effects of the lot sizing on inventories in multilevel production systems while following a reorder-point policy (cf. Orlicky 1975, p. 26)

More recent analyses based on SCM show that isolated planning with MRP systems (which ignores the interrelationships in the SC) also results in unsatisfactory overall system behavior, and coordination of isolated MRP systems by SCM should therefore lead to considerable improvements.

MRP systems support the determination of production lots and ordering quantities by providing different heuristics. The classic inventory models are based on rather restrictive assumptions. Compared with the traditional theory, the best-known representative of which is the Harris-Wilson formula, the procedures for determining lot sizes in MRP systems differ primarily in allowing time-dependent demands. A multi-period BOM explosion often leads to significantly different secondary demands that cause unstable work loads and material flows.

R/3 provides various heuristics for material requirements planning, in the form of static (e.g., lot-for-lot, fixed lot size) and dynamic methods (e.g., least-unit cost, part period balancing, and Groff's lot sizing procedure).

The capabilities of the different heuristics used to determine lot sizes have often been compared. The Groff and Silver/Meal procedures give the best results when there are positive demands in all periods. However, these heuristics can be significantly improved for demand patterns in which no demands exist in many periods (e.g., because of the effects of lot sizing in the higher levels of the part hierarchy) (Knolmayer 1987b).

From a logistics perspective, the aforementioned heuristics are sometimes criticized because the cost parameters needed for calculating order quantities or lot sizes can seldom be determined exactly and the small cost advantages that result from lot sizing, especially when setup times are short, may be more than compensated by the irregular and badly plannable material flows and capacity demands. Rather than inventory optimization, the IS should endeavor to provide a smoothened material flow (Knolmayer 1987a). More regular demand patterns are also obtained if the bullwhip effect can be reduced, for example by collaborative forecasting and improved information exchange within the SC (cf. section 3.1.2).

IBM developed an Asset Management Tool (AMT) as an SC analysis tool for internal use (Lin et al. 2000, pp. 15). The tool embodies a coupling of (so-called) optimization, performance evaluation, and simulation, integrated with data connectivity and an Internet-enabled modeling framework. The user can perform very deep what-if-analyses that are beyond the capability of standard simulation tools. The simulation engine can be used to induce the inventory module to perform periodic recalculations of inventory levels while simulating dynamic SC policies and processes. SC are modeled as networks of inventory queues, using a decomposition scheme and queuing analysis to capture the performance of each stocking location. The user may validate and fine-tune SC parameters based on analytical results. IBM claims that its Personal Systems Group saved more than $ 750 million in 1998 by using the AMT applications.

Production Control

Production control follows production planning activities and is concerned only with those orders that are to be processed in the near future. Its main tasks are determining the time to release an order, resource scheduling, and sequencing of production orders.

The order release passes a production order to the shop floor. The timing of this has a significant effect on how far the main goals of production management are achieved: Inventory levels, lead-times, meeting delivery dates, and capacity utilization rates are highly dependent on the order release date (Wiendahl 1995, pp. 203). Many priority rules have been developed and compared experimentally for sequencing; these include, First Come First Served, shortest operation time, longest operation time, and slack time rules. The comparisons sometimes yield contradictory results; these may be caused by assumptions about production processes and capacity utilization rates.

In production control, the resource assignments of lead-time scheduling must often be stated more precisely and modified using current data from the shop floor. For example, during lead-time scheduling the MRP system assigns the earliest and latest time points for starting and finishing each operation. These data are extended during detailed planning, the exact actual start and end times being inserted. This requires that in the data models entities and relationships must comprehend additional attributes; thus, the entity type for operations continues to be used but new attributes and new attribute values are assigned to it (cf. Becker/Rosemann 1993, p. 103).

On the shop floor many unexpected events occur; therefore assignments and schedules must be sometimes revised. Often the new schedule is not determined from scratch but the previous one is "repaired" in some sense ("reactive scheduling"; "repair-based scheduling"). The schedule may be improved by a local modification (e.g., by exchanging two jobs). However, a sequence of minor changes will often stop at a local optimum. Therefore, broader techniques like tabu search, simulated annealing, and genetic algorithms have been applied for reactive scheduling (Dorn et al. 1998, pp. 280).

Some MRP systems exclude the detailed scheduling and realization phases from integrated order management, for example by using conventional or electronic control units or self-controlling feedback cycles (Kanban system). Potential developments are envisaged in the multimedia content of electronic control units and the increased use of artificial intelligence methods (Kurbel 1992; Kurbel 1993).

The functionality SAP previously provided for its mainframe product R/2 as a separate electronic leitstand has been integrated in R/3. It is now assigned to the PP/DS module of the APO (cf. section 3.1.3).

Correct and current feedback messages (plant data collection) are of major importance for production control. The error rates for human inputs can be significantly reduced by using barcodes, transponders, and other electronic data capturing techniques (cf. section 3.3). If necessary, alternative routings may be prepared and used to handle unforeseen situations. System-supported, knowledge-based recommendations can be envisaged in this context. Furthermore, manufacturing orders may be overlapped or split to reduce the throughput times; however, this results in additional coordination and setup tasks. Customer changes, corrective work, and repair orders must also be taken into account.

Many companies attempt to compensate out-dated MRP results by employing "progress chasers" who try to increment the priority figures for delayed production orders. Such interventions may have undesirable effects on other orders and worsen the ability to plan and control the production processes.

2.4.3 Non-traditional Concepts

The interest in competitive aspects of production management ("manufacturing matters"; cf. Cohen/Zysman 1987) has given rise to a number of concepts, some of which are at odds with the mainstream MRP logic and have become the subject of controversial discussions. These concepts include, in particular, Kanban and JiT systems, OPT, and cumulative quantity systems; we describe these methods insofar as they are relevant for SCM.

Kanban and Just-in-Time

The Kanban system was developed as long ago as in the 1950s and is the most important component of the "Toyota Production System" (Monden 1993). Since the 1980s, Kanban has also attracted significant attention in the US and in Europe (Wildemann 1985; Sakakibara et al. 1993; White et al. 1999). When a Kanban system is used in a Western company, often only a limited part of the plant is organized in this manner ("Kanban islands"). The JiT systems developed as a response to the Japanese Kanban concept differ from it primarily in the intensive use of information and telecommunications systems.

Kanbans are cards that are placed on standardized containers. The circulation of the cards between the different echelons of the SC results in a system with self-controlled feedback loops. As soon as consumption of the parts in a certain container begins, the card is removed from it and transported to the previous element of the chain. The arrival of the card indicates that there will soon be a new requirement to deliver parts ("supermarket principle"). Buffer storage with a minimum inventory level is defined for every element. Its size depends on the number of Kanbans in circulation; several methods have been proposed for determining these numbers (Mak/Wong 1999; Alabas et al. 2000). The resulting feedback loops are intended both to reduce the inventories on the shop floor and to shorten the lead-times. Important preconditions for the usability of the Kanban system are:

- Material-flow-oriented layout of resources and workplaces
- Sales policy measures aimed at stabilizing the sales and, thus, the production program
- Significant flexibility of the workers (with regard to the type of work performed and to working hours)
- High availability of the resources
- Provision of backup machines or external capacity
- Short setup times for the resources
- Small lot sizes, determined by container size
- Sophisticated quality assurance concepts.

Unstable demands lead to additional buffer inventories (Takahashi/Nakamura 1999). Thus, reducing the bullwhip effect should also mean smaller safety stocks are needed in Kanban environments.

A Kanban system does not necessarily need computer support. This is not surprising, because computers were not available for business applications when the Kanban system originated in the 1950s. Coordination problems can arise, in particular, if some production processes are controlled by MRP systems and others by Kanban principles. To avoid such problems, Kanban-controlled parts can also be included in the ERP system and the Kanban mechanisms be reproduced in the IS. In such concepts, reading of barcodes replaces the physical movement of cards. For conventional Kanban systems, an event in the subsequent element of a logistical chain always results in actions in this or in the immediately preceding element. IS can extend this event orientation over all elements of the SC. From this viewpoint, Kanban realizes an event-action mechanism which is a special case of action-oriented data processing and business-rule-based IS (cf. Herbst 1997).

The R/3 PP module supports the Kanban principle. Objects are cards, production supply areas, control cycles, and Kanban boards that provide an overview of the current status (e.g., full, empty, in progress, being transported) of the containers. The board visualizes bottlenecks and problems arising with material supply. The R/3 Kanban system allows external procurement, internal production, and supply from a warehouse. The initiating event for delivery of the material is a status change at the container; when the container status is altered from "full" to "empty", R/3 automatically generates replacement bookings. The status of the containers can be changed directly on the Kanban board, in an input mask, or by scanning a barcode printed on the card. When a receiver sets the status of a container to empty, a replacement element is created and the associated source is requested to supply the material. As soon as the status changes to "full", the arrival of the material is booked automatically with reference to the procurement element. A supplier can view the inventory levels of materials via the Internet and determine what quantities of the materials need to be provided. It can define a delivery due list and inform the customer by setting the status at "in progress".

If ERP systems are used, a reliable sales forecast permits parts to be available in case they are needed ("Just-in-Case"). This results in a type of inventory that some management philosophies condemn as being the root of all evil: Competitive production should be sufficiently flexible to react very fast to requirements that arise externally or internally and, thus, able to respond "just-in-time". JiT principles mean smaller stocks; the financial resources freed up by reducing the inventory can be used to extend the flexibility of production systems.

JiT deliveries are based on close and coordination-intensive cooperation; therefore the implementation of JiT relationships often requires a reduction in the number of suppliers (cf. section 2.3.1). If the parts have to be delivered directly to a customer's conveyor belts without any intermediate storage, the two companies concerned must cooperate very closely in areas such as quality assurance. The application preconditions for JiT logistics (Copacino 1997, pp. 58) largely correspond to those formulated above for the Kanban system.

Because inter-plant JiT deliveries normally use road transport, they are undesirable from an environmental point of view and liable to delays (danger of being "Just-in-Traffic jam"). Sometimes logistics service providers maintain forwarding warehouses close to the plant to permit JiT deliveries. Some companies offer consulting services for their suppliers, to help them in building up or improving a JiT system. However, the concepts developed in the automotive industry, for example, cannot be transferred to other industries without considering differences in the applied technologies. Thus, process interruptions are conceivable in an SC in which a favored JiT delivery can only be provided from a warehouse. Significant advantages of applying the JiT concept are only achieved if it can be implemented along the whole of the SC.

Bottleneck-oriented shop floor control

This procedure attempts to identify bottlenecks and to use this knowledge in production control. The best known procedure is OPT. Although the method is described in many publications, it has seldom been implemented as an independent software solution.

Recently, SCM systems contain bottleneck-oriented procedures for production control ("constraint-based scheduling"). From the functionality point of view such components should be part of the production planning modules of ERP systems.

Various (mostly nine or ten) basic rules for shop floor control have been formulated in conjunction with OPT (cf. Jones/Roberts 1990, pp. 37):

1. Balance flow not capacity
2. The level of utilization of a non-bottleneck resource is not determined by its own potential but some constraints in the system
3. Utilization and activation of a resource are not synonymous
4. An hour lost at a bottleneck is an hour lost for the total system

5. An hour saved at a non-bottleneck is just a mirage
6. Bottlenecks govern both throughput and inventories
7. The transfer batch many times should not be equal the process batch
8. The process batch should be variable not fixed
9. Schedules should be established by looking at all of the constraints simultaneously. Lead-times are the result of a schedule and cannot be predetermined.

The often-quoted higher level statement "The sum of the local optima is not equal to the global optimum" is mentioned in connection with OPT as well as SCM.

Whereas Kanban and other methods, e.g., the load-oriented order release (Wiendahl 1995, pp. 203), reflect attempts to reduce the inventory levels throughout the whole system, OPT recommends differentiated policies in which inventories are concentrated at its strategic points. It would be extremely undesirable if the production at a bottleneck resource is interrupted because no orders were waiting at a particular time; this situation could be caused by failures in previous manufacturing levels. In keeping with the principle of "protecting the bottlenecks", safety stocks held in front of the bottlenecks guarantee with sufficient probability that an interruption such as that described above will not occur.

Conventional planning philosophies deplore the existence of bottlenecks and define their elimination as an important strategic goal. If bottlenecks are identified by OPT they should remain stable in the medium term, because planning and scheduling procedures can then be concentrated on specific groups of production facilities.

SAP APO provides bottleneck-oriented procedures (cf. section 3.1) by mathematical methods implemented on the basis of „Optimization engines" from ILOG [http://www.ilog.com/]. This company is the leading vendor of mathematical components for SCM software. There is also a model library from which the user may generate customized models that take characteristics of the particular industry concerned into account. The user can define for each resource group separately whether or not it has to be considered in the "finite planning" procedures that take capacity constraints into account.

The OPT software uses dynamic simulation models to prepare production plans. Towill (1996) provides details on possible uses of simulation models to design the SC. The importance of considering stochastic phenomena in SCM too leads to a hierarchical planning approach in which optimization methods and heuristics are supplemented by simulation models (Kuhn/Werner 2000).

The Supply Net Management Research Group at DaimlerChrysler modeled, simulated, and analyzed a four-stage, cross-company SC for parts delivery as part of a larger collaborative planning pilot project that explored the potential of real-time information exchange with its suppliers. The results of the simulations were used to evaluate the effects of particular planning policies on performance and stability of the SC in term of oscillations, inventory levels, stock-outs, and logistics service (Baumgaertel et al. 2001).

SAP is planning to enlarge its APS system with more powerful simulation capabilities in forthcoming releases.

Cumulative Quantities

Cumulative quantities are the quanta of certain parts that have been produced, called up or dispatched since some specified starting date (Mertens 2000, pp. 102). They are visualized in a coordinate system. Comparison of the cumulative quantities yields initial information about the synchronization of demand and supply for successive elements of an SC. The cumulative quantities can also be used within the company in a distributed production system. If the concept is used over s stages, s-1 feedback loops result: As with the Kanban system, only adjacent elements in the SC communicate with each other. The provision of cumulative quantities for all members of an SC in an Information Warehouse would significantly improve the coordination capabilities.

2.5 Distribution

Whereas determining the sales outlet routes is one of the sales department's responsibilities, the physical distribution involves all packaging, warehousing, and transportation activities needed for the products to reach the customers. The term "marketing logistics" is sometimes used as a synonym for physical distribution. In this section we discuss how physical distribution may be reorganized to improve SCM.

Distribution planning is characterized by (cf. Geoffrion/Powers 1995, p. 107)

- an astonishing variety of real management issues marching under a banner originally reading "warehouse location",
- an amazing longevity of technically inferior, non-optimizing solution methods,
- a lack of new algorithmic breakthroughs since the mid-1970s, and
- the huge effort required for data modeling compared to that needed for the implementation of optimization methods.

The affiliation of distribution planning with SCM systems will be very helpful in the context mentioned above. Warehousing and transportation have been identified as functions with large cost saving potential (Baumgarten/Wolff 1999, p. 59).

A study by Automation Research Corp. showed that SC for which distribution is of major importance are more ready to invest in APS systems (refer to sections

2.4.2 and 3.1.3) than companies that have to focus on procurement or production issues (NN 1998).

2.5.1 Distribution Types and Tasks

The time perspective of planning influences the degree of freedom that can be applied in determining the distribution channels. In long-term planning, the company must decide whether, and possibly where, plants and central, regional and/or customer warehouses have to be established and what policies should be adopted at the individual locations. Figure 2.11 illustrates some options for determining distribution networks.

The decision on the type of the distribution system is influenced by the requirements that customers have in terms of service levels and delivery times and by the costs associated with the various distribution strategies. In particular, the choice of locations for warehouses can be supported by mixed-integer programming models (Hummeltenberg 1981).

The integration of the European markets resulted in a tendency to centralize warehousing (cf. Kobler 1997, pp. 87; van Hoek 1998; Alt et al. 2000); this can be attributed to such developments as the standardization of registration procedures and product specifications.

Nike had approximately 25 warehouses in Europe in 1993; five years later it was operating only one European distribution center, in Belgium (Ashford 1997).

Ciba Performance Polymers has used SAP R/3 to restructure 13 logistics systems with 26 warehousing locations in Europe to a single logistics organization with only six regional distribution centers.

Traditionally, customers are supplied either from a central warehouse (source-based distribution) or from local warehouses (market-oriented distribution). "Cross-docking distribution" (cf. Figure 2.12) tries to combine the advantages of both organizational forms. Although the goods are brought to regional distribution centers or to a central warehouse belonging to the retailer, they are meant to pass through as fast as possible; such goods have already been assigned to specific recipients and, if necessary, may be picked or relabeled for specific branches. Typically, the goods are transported to the customer on the day after arrival. A precondition for the functioning of cross-docking is to synchronize receipt and issuing of goods. In the cross-docking approach, advanced IS replace the functions of the warehouse stocks.

The cross-docking concept has been developed at Wal-Mart. Applying the hub-and-spoke principle designed for determining networks of aircraft routes, every warehouse in the US is responsible for approximately 175 branches within a radius of 150 - 300 miles. Each transport supplies two to three branches. The warehouses use laser-controlled transportation systems where barcodes indicate to which trucks the incoming goods have to be routed. On average, the goods do not spend more than 48 hours in a warehouse (Nicholas 1998, pp. 697).

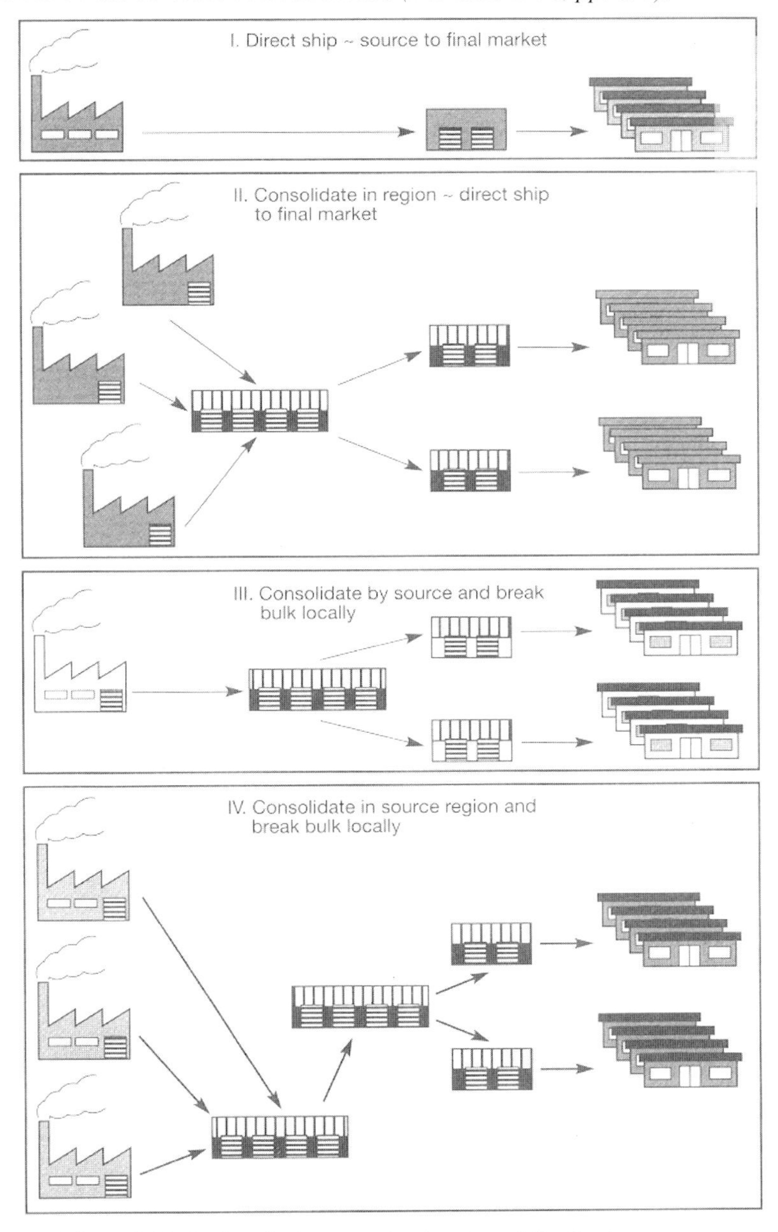

Figure 2.11: Options in the strategic planning of distribution systems (Christopher 1998, p. 139)

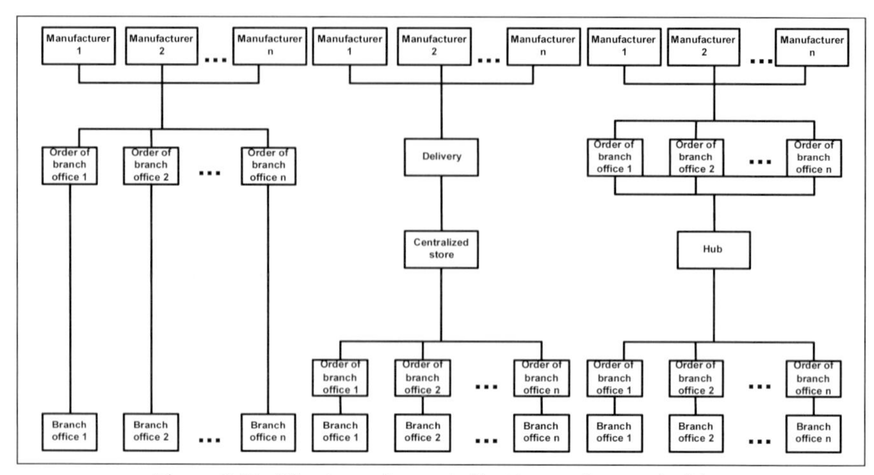

Figure 2.12: Direct supply, centralization, and cross-docking (cf. Seifert 1999, p. 93)

The postponement concept of product design (cf. section 2.1.3) is relevant in two ways for the role of distribution centers in an SC. On the one hand, the concept can be applied to the distribution system so that the spare parts for which demand arises randomly and is thus scarcely predictable, in particular with regard to the demand region (slow movers), are stored only centrally. Fast and cost-effective transportation systems and efficient customs handling procedures are essential preconditions for this distribution strategy, which avoids the risks of decentralized warehousing. Only those spare parts for which the demand is easy to forecast (fast movers) are stored in the decentralized storage facilities. On the other hand, attempts are made to shift the postponement into the distribution channel (van Hoek 1998).

Colorado Cyclist, a mail-order retailer, sells bicycles and allows the customer to select a frame along with any desired components (down to the last spoke nipple). Thus, the boundaries between manufacturer und retailer vanish and new forms of cooperation arise in an SC (Ulrich et al. 1998, p. 196).

To avoid time-consuming and stressful priority discussions, the spare parts warehouses are usually separate from the factory warehouses not only in terms of responsibility but also physically. This policy is realized because the plant locations are often unsuitable for distribution purposes and the central warehouse is established at a location that is better suited for distribution. In addition, handling of spare parts requires a different picking technique than that required for production purposes.

The concept of manufacturing segmentation (cf. section 2.4.1) can be extended in such a way that, to reduce the number of interfaces, the responsibility for distribution is also decentralized to the segments.

From the SC viewpoint, the coordination between production scheduling and distribution is of special interest. In many companies, organization and incentive structures are directed toward improvements within manufacturing or distribution. These efforts have often reached a zone of diminishing returns, so that one should concentrate on achieving better coordination between production and distribution. Complex heuristics have been proposed for attaining this goal (Chandra/Fisher 1994). Experimental results show that the value of coordination can be remarkably high; this holds especially if the production capacity constraints are loose, because in this case there are more options for improving the distribution schedule by means of consolidation. As may be assumed, the value of coordination also rises when the distribution costs are high and when there are comparatively large numbers of products or customer locations.

2.5.2 Intermediaries in the Distribution Process

While the SC is being designed it is necessary to decide whether and how intermediaries are to be used. Because establishing an in-house transportation capacity for irregular deliveries to customers is often uneconomic, there is a long tradition of outsourcing transportation tasks to external carriers (Razzaque/Sheng 1998). A study conducted by PE Consulting investigated the satisfaction levels of companies that have outsourced their distribution tasks. Although approximately two-thirds of the companies had achieved their goals with outsourcing, about ten percent were very disappointed with the results of this policy (Sturrock 1999). According to a 1995 survey, the most frequently cited benefits of using external distribution services include

- lower cost (38%),
- improved expertise/market knowledge and data access (24%),
- improved operational efficiency (11%),
- improved customer service (9%),
- ability to focus on core business (7%), and
- greater flexibility (5%) (Dornier et al. 1998).

The core products of various providers of logistics services are often rather simple and easy to substitute. Consequently, many carriers attempt to extend and differentiate their services to obtain competitive advantages. Those offerings outside the core activities are known as value-added services. Examples are:

- *Barcoding:* The manufacturer receives from the carrier barcode labels that can be placed on the goods to be transported. The carrier scans the shipping goods at various "checkpoints"; the information recorded when these points are passed is made accessible to the customer (e.g., Federal Express data can be accessed by customers via http://www.fedex.com/us/tracking/).
- *Warehousing:* Various carriers offer the use of regional distribution warehouses. For example, the Swiss Federal Railway Company has established

different warehouses specializing in the requirements of specific customer groups.

- *Extensive service network:* An extensive network of branches is of particular importance for goods with short lead-times and small delivery quantities. This applies, for example, to convenience goods.
- *Spare parts services:* Some carriers offer a supplementary service of direct delivery of spare parts within a few hours once the requirement is made known.
- *Reusable material logistics:* This service returns packaging material or transportation equipment for reuse. However, attempts to avoid returning of transportation equipment are intensifying; this also implies, for example, modified behavior among the consumers (e.g., acceptance and recycling of PET instead of glass bottles).

The standardization of pallets (e.g., by adoption of "Euro pallets") obviates the need for returning empty transport equipment to the shipping point; such return shipments should be used only for peak quantities, whereas the majority of pallets can be used by companies located close to the destination point.

Hewlett-Packard in 1995 passed the responsibility for all logistical handling, of the products manufactured in Singapore and required in Europe, including responsibility for data collection, to the German shipper Kühne & Nagel. This applied to data relating to sales forecasts, new orders, warehouse inventory, production schedules, and transport. Thus, the shipper provided value-added services for planning of sales, production and transport planning, and for warehouse control. Because of the resulting dependencies, this extensive range of tasks transferred was later reduced (Boutellier/Locker 1998, p. 87).

Nowadays, the concept of third-party logistics, which provides for outsourcing of distribution tasks (Buxmann/König 2000, p. 52), is being countered by the vision of fourth party logistics (4PL) (Figure 2.13). It is argued that the shippers (as "third party") do not have the necessary know-how or the independence to identify the currently best offerer of transportation and warehousing services and to provide the integration services that are needed to meet the customers' requirements. Consultancies see themselves as providers of 4PL services (Foster 1999) and therefore wish to establish themselves as "fourth parties".

The attempt of many manufacturers to change from traditional sales and distribution channels to direct contact with the consumers can be seen as a policy of elimination of intermediaries and a counter-movement aimed at including ever more service providers. A well-known example is the direct sales of hardware, as offered by Dell Computers or Cisco Systems over the Internet (Christopher 1998, pp. 19). This distribution channel permits the customer behavior to be perceived more directly than if intermediaries participate in the SC; on the other hand, some suppliers are afraid of "cannibalization effects" in the traditional distribution channels that are usually still far more important than EC with respect to the volumes concerned. If the customer contact is established via the Internet, attempts must be made to ensure that the customer's wishes can be satisfied without procedures needing additional clarification; this requires observation not of only simple integrity conditions (e.g., "error-in-time sequence" for travel and transfer times in air transport) but possibly also comprehensive configuration aids (e.g., for the configuration of PCs). Furthermore, database access is needed, for example, to check the availability of goods that are on-hand only in limited quantities against the requested order quantity before confirmation.

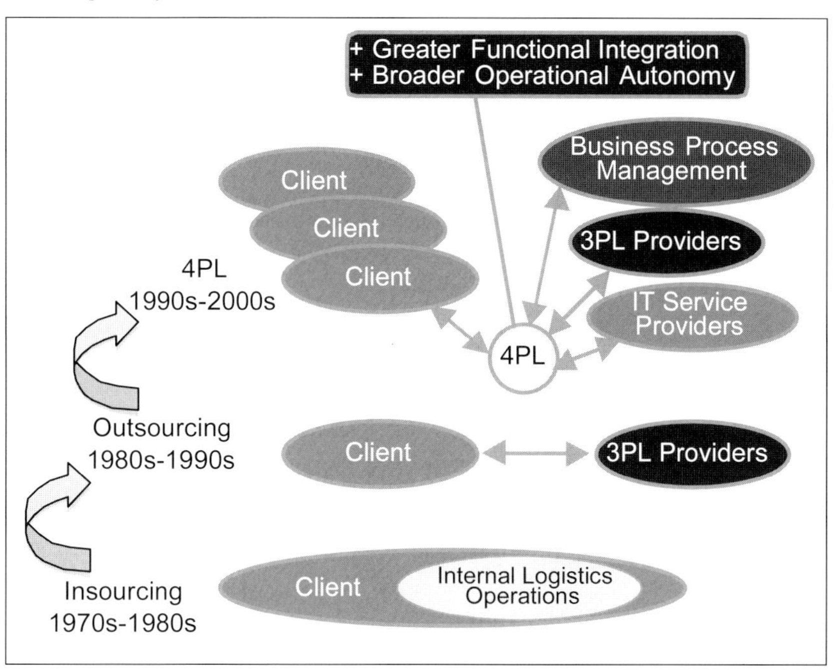

Figure 2.13: Development tendencies in organizing the distribution from the viewpoint of the former Andersen Consulting

2.5.3 Information Systems in Distribution

In conventional distribution systems the transportation documents are printed and the printouts tied in with the physical movements. IS uncouple the information flows from the underlying material flows and can thus improve the transportation processes in the SC.

Optimization models

Optimization models that support distribution decisions depend on (cf. Daganzo 1999)

- whether the goods to be transported are made available at one or more locations and whether they are be delivered to one or more locations (one-to-one, one-to-many, many-to-many distribution) and
- how many levels, and thus transshipments, the distribution system covers (factory warehouse, central warehouse, regional warehouse, etc.).

There is a huge body of literature proposing the use of mixed-integer programming models for supporting strategic production-distribution decisions. A detailed tabular overview of the main features of eight highly relevant papers is provided by Vidal/Goetschalckx (1997, pp. 10). Factors considered in some models are disregarded in others; thus, sets of inter-related submodels as parts of a general decision-support system for use in designing SC would be helpful. Each component should focus on representing several factors and be flexible enough to interrelate with other components. A "logistics laboratory" has to include optimization models, heuristics, and simulation capabilities (Slats et al. 1995).

Potential pitfalls of modeling global SC are

- performance measures that are unsuitable from SC viewpoints,
- inadequate definition of customer service,
- inappropriate aggregation of customers to demand clusters,
- excluding uncertainties,
- disregard of international aspects, such as exchange rates, taxes and duties, transfer prices, infrastructures, reliability of transportation channels, and trade barriers,
- neglect of coordination issues,
- consideration of only partial aspects of shipment methods,
- incorrect treatment of inventory costs, and
- separation of SC design from operational decisions (Lee/Billington 1992; Vidal/Goetschalckx 1997).

As always, the real system can be modeled with different degrees of accuracy. Mid- and short-term decisions must be made on which products, from which locations, have to be delivered to which customers, and in what sequence this should be done. Although, from a theoretical point of view, these problems are interrelated and should be treated simultaneously, their complexity usually leads to stepwise planning procedures.

Transport models

The classic transport model assumes that specified quantities are available and have been requested at particular locations. The simplex method or specialized algorithms that make use of the formal structure of the transportation problem can be applied to solve this type of linear programming problem. Because determining the actual distances between many points is complex, simplifications for measuring distances are often applied (cf. Bramel/Simchi-Levi 1997, pp. 259). Even if the distances are known exactly, the topography and the (traffic) conditions in the agglomerations are normally ignored. The costs for traveling certain routes can vary, for instance, depending on the time of the day and on weather conditions.

Route planning models

After assigning the delivery orders to specific distributing warehouses, in particular in the case of road distribution, the problem of the sequence in which the selected transport vehicles should supply the demand points arises (route planning). These sequences can be specified by route planners or determined by the drivers in accordance with constraints resulting from customer agreements. What is sought is a round trip that satisfies specific requirements, for example, with regard to particular receiving intervals, maximum travel times, and maximum distances. Traveling salesman models have been developed to determine cost-minimizing round trips. When a large number of locations has to be served, the resulting magnitude of combinations means that such models can seldom be solved exactly. Also the practical restrictions mentioned above can scarcely be considered. For this reason, most software packages use heuristics for route planning; sometimes expert systems are applied. In addition to determining the travel routes, the route planning systems can perform additional tasks:

- Graphical display of travel routes on a screen
- Output of route lists, customer lists, and order lists
- Instructions for the drivers to simplify finding the exact location of the destinations, nowadays also available as display in the driver's cab
- Checking the economics by calculating vehicle-loading statistics and transportation costs.

Table 2.3 summarizes potential advantages of using route optimization programs.

Table 2.3: Advantages of route optimization programs (cf. SoftGuide 2001)

Advantages to the customer	• Improvement of customer service • Increased reliability through monitoring delivery dates • Competitive advantages resulting from reduced delivery times • Quick response to special requests
Advantages to management	• Increased transparency • Independence from the planner's intuition • Simpler training of new employees • Reliable data for decisions on the transportation fleet
Advantages to the scheduler	• Reduction of routine tasks • No misplanning resulting from typing or calculation errors
Advantages to the transportation fleet	• Independence of traditional route plans • Higher loading factors • Reduction of the number of unbilled journeys

The SAP Logistics Execution System (LES; cf. section 3.3) is one component of the mySAP SCM applications. Route planning is initialized in LES by selecting the means of transport; the resulting routes are determined by the system using previously defined rules. An optimization in the traditional sense (e.g., selection of the lowest-cost route) is not provided by LES. However, interfaces can be used to connect route optimization programs from certified SAP partners with the LES. Furthermore, additional functionality for transportation planning and vehicle scheduling (TP/VS) is available from APO Release 3.0 upwards (cf. section 3.1.4).

Distribution Requirements Planning Systems

Traditionally, the suppliers respond to an order but not to the changes in the warehouse inventory in the distribution network, which remain hidden to them. If there is a sudden demand increase, this may result in orders that cannot be fulfilled promptly. Moreover, reduced demand will often lead to an increased warehouse inventory. Distribution Requirements Planning (DRP) systems are related to MRP systems and determine time-related schedules for moving goods through the distribution network (cf. Toomey 1996, pp. 112). They are particularly suitable for supporting multi-level distribution systems.

Many companies have reduced the number of levels in their distribution systems over the last few decades (cf. section 2.5.1). The simplification of the real system (in a similar way to that seen in manufacturing, cf. sections 1.2.3 and 2.4.1) reduced the need for complex planning and scheduling software. Steps towards SCM have resulted in "Channel DRP systems" to improve the information flow in the logistics chain and thus provide better planning data (Copacino 1997, pp. 102).

Carriers often do not use packaged software and thus lack the integrative capabilities of ERP systems. Only approximately one third of the polled US carriers use such systems as SAP R/3, PeopleSoft or Oracle Applications (NN 1999d).

Distribution control stations

Distribution control stations are similar to the control stations ("leitstands") in the manufacturing department (cf. section 2.4.2). Requirements for these control stations are, in particular (cf. Mertens 2000, pp. 180):

- Graphical user interface with display and intervention capabilities similar to the planning board of a manufacturing control station
- Method base for route planning
- IS supporting drivers and the assignment of vehicles
- Visualization of exceptional situations (e.g., flashing signals for delays)
- Connections to vehicle locator systems.

The use of a distribution control station permits a scheduler to plan urgent deliveries flexibly in real-time, to modify routes as the result of unforeseen situations, and to respond to disruptions in the delivery. Route optimization models can be made accessible via these stations.

Fleet management systems and shipment tracking systems

Linking the vehicle personnel to the company's IS permits real-time tracking of the vehicles and dynamic route planning. The resulting plans can be updated centrally by a vehicle scheduler or in a decentralized manner by the truckers. This requires that the cabs of the trucks are fitted with in-vehicle computer systems that allow sending and receiving messages (Mertens 2000, pp. 234).

It should be possible to link the order data with the customer and purchasing data during manufacturing and distribution. Customers are increasingly offered the capability to track the status of their order online. This information is either requested by the customer or automatically sent to the E-mail address specified when the order was placed.

The customer is automatically given an order number when an order is processed by the SAP Online Store. This number can be used to assign the order status in all subsequent information requests.

To determine vehicle positions, fleet management systems are often linked with a Global Positioning System (GPS). This system, based on satellite technology, can be used to determine the exact location of a vehicle to approximately 150 ft and thus provides customers with information about the whereabouts of the good in transit and its expected time of arrival. It can also be used, for example, to locate stolen vehicles.

Shipment tracking systems are also known as "track and trace" systems. Problems arise with media fragmentation; for example, not all airline companies offer shipment tracking. SAP offers an SCM Event Management solution that also encompasses a Tracking&Tracing Server.

Federal Express and SAP have jointly developed software that automatically generates the shipping documents and tracking numbers that FedEx needs for shipment tracking; the R/3 database is updated as soon as the product has been delivered (Sweat 1998).

ECOutlook.com [http://www.ecoutlook.com/] offers a track-and-trace system that provides access to the tracking systems of all major carriers via a single point of contact.

Transport exchanges

Computer-supported transport exchanges supply information about the availability of freight and loading space with regard to both geographic and time aspects. The main goal is an improved use of existing transport capacity, in particular a reduction of the "dead" freight volume. Transport exchanges are gaining in importance with the wide use of the Internet.

Most transport exchanges operate as selection systems (such as electronic bulletin boards); electronic matching of supply and demand does not (yet) take place. They differ in the geographic catchment area and the range of services offered. Whereas some exchanges restrict themselves to the brokerage of freight space for road transport, others also procure the services of shipping and airline companies and/or offer the use of online route planning programs, cost calculations, and logistics consulting. Depending on the provider, users of a transport exchange may be liable for payment.

SAP LES contains the function "tendering for bids" by which planners can offer shipments to carriers via the Internet. They can react to the offers made by carriers and also supervise the status of the tenders. All interaction between the planner and the service agent goes through the system. Planners select the carriers to whom they wish to offer the shipments, set the tender status, and specify conditions for shipment processing. The planners see the tender status and receive an overview of the most important information on the shipment tender. They can also include carriers that do not use an SAP system. Access is monitored using safety profiles and authorization objects that are assigned specifically for the tender status tasks. The service agents can call up only those shipments that were directed to them.

The dedicated monitor gives an overview of the tendering status for each transportation planning point. The number of shipments is listed by status. In this way, the transportation planner can respond quickly to definite jobs on offer and also obtain a complete overview of the overall status of all shipments listed for tendering, any of which can be displayed if no answer has been received from the forwarding agent within a specified interval.

Carriers can call up the planner's Internet page and react to the offers listed there. They can view the shipments offered and accept them as offered, accept them with added conditions, or reject them.

The system displays the processing status of the accepted shipment orders. Carriers plan their transportation resources and update the actual dates for the beginning and end of loading, from which the duration of the activity can be derived.

2.6 Service Management

Customer-oriented services are gaining in importance over the technical properties of the products in long-term customer relationships. Consequently, the service management group is important for the retention and expansion of SC partnerships.

In SAP R/3, the Service Management (SM) module provides the following functions:

- *Administration of customer objects*
- *Warranty check*
- *Administration of service contracts*
- *Recording of service requests*
- *Processing of service orders*
- *Supply of spare parts*
- *Invoicing*
- *Reporting (cf. Buxmann/König 2000, pp. 74).*

In the following section we consider only the service activities that supplement order management in manufacturing plants.

2.6.1 Call Center

A call center employs staff who provides, among others services, telephone support for the company's customers. Call centers are often used mainly in the after-sales phase, when, for example, customers require explanations about how to use the product they have purchased or complain about malfunctions. A call center might have the goal of solving, for example, at least 80% of all problems directly by telephone.

A call center system usually consists of modules for recording problems, for activity control, and for fault diagnosis and repair (problem management systems). Typical characteristics of such solutions are:

1. Calls go to a group of processing staff and not to individual persons
2. A special communication system is used to distribute calls to the processing staff
3. All contacts are documented in order to allow monitoring and evaluation of the status and time needed for solving the problems.

An important aid for call centers is Computer Telephony Integration (CTI) which supports initiating, receiving, and handling of telephone calls by the use of IT. The first step in processing a customer query involves using predefined dialogs to solve the problem or pass the customer to an appropriate employee. Care should be taken to ensure that customers are not annoyed by long-winded or poorly-structured dialogs. An internal goal of Automated Call Directing is an approximately uniform distribution of calls to the agents.

Typically, call centers are organized according to business processes (e.g., complaints management). The staff members advise callers, process complaints, or transfer the calls when a contact partner is requested by name or when specialized knowledge is needed to solve the problem. The staff can query databases for quick access to the current data. Such information might be, for example, products previously purchased by the customer, payments still outstanding, services that have been provided to the caller or faults that may arise with a certain product and suitable means of correcting them.

Escalation procedures are often defined: If a problem cannot be solved within a specified time interval, the call is passed to a colleague who is assumed to be more competent; at the same time, a message may also be sent to the department manager about this problem situation. Archiving (cf. section 2.6.3) and analyses of the problems by specialized tools play an important role. First, it can prove possible to use previous cases to solve current problems; secondly, analyses can show up strengths and weaknesses of the support mechanisms and make it obvious when reorganization is needed.

SAP has extended R/3 with the SAPphone interface and the front office Customer Interaction Center (SAP CIC) to provide relevant data in the call center and to simplify the integration with call center systems from third parties. SAP and Siemens offer joint solutions. They range from the recognition of the callers based on the telephone number, include the automatic start of an R/3 transaction (e.g., displaying open orders of the customer) with preassigned caller data, the initiation of checkbacks, conference switching, and call forwarding.

2.6.2 Help Desk

A help desk supports staff especially in dealing with difficulties in using hard- and software. The extent of help desk utilization depends on the amount of training received by the employees. User support is often organized in several levels. Because of the similarity of the tasks performed, the call center and the help desk are merged in some companies. Outsourcing of help desk tasks to external service providers may be an economically sound option (Mertens/Knolmayer 1998, pp. 22). When an enterprise is integrated into an SC, its staff can be given access to a help desk that was originally set up for internal support within the company that started it.

A help desk system can be organized using the SAP Call Center components. The R/3 System and the Siemens products for CTI provide the required functions.

2.6.3 Archiving Systems

Archiving systems are used for long-term storage of documents in an unchangeable, digitalized form and for quick retrieval. There are archiving systems that integrate a database and those that use a database. Archiving systems help to save personnel costs for sorting and searching and make the documents stored easily available to all authorized employees in the company. The archived data may also be accessed by staff of other companies in an SC. If a retailer needs a document from the development department of the manufacturer, transport time and cost can be reduced if it accesses the electronic document via an IS. Documents that are relevant for several partners in the SC may be filed in a centralized archiving system.

In the R/3 environment, archiving is understood as moving data from the online system to offline storage. The Archive Development Kit is part of the R/3 development platform and supports the creation of archiving and query programs. It includes archiving classes that cover a set of function modules for a multiple-archived business object; this permits change documents to be parts of accounting or sales documents, for example (Schaarschmidt/Röder 1997).

SAP offers the capability of linking archiving systems from third-parties to R/3 Systems using the certified standard interface SAP ArchiveLink. In particular, SAP cooperates closely with IXOS, the market leader for archiving applications in SAP environments. The IXOS-ARCHIVE software provides relevant business documents, such as incoming invoices, E-mails, fax messages or spreadsheets for R/3 business processes.

2.7 Recycling and Disposal

At the end of their useful life, all products purchased must be reused, possibly in modified form, by a member of the SC or another company or be disposed. For ecological reasons, attempts are increasingly being made to avoid dumping products (e.g., on garbage dumps), the preference being to reuse them in whole or in part (recycling). This also applies to waste, rejects, and by-products. Thus, the delivery network has to be extended by a recycling and disposal network (Reverse Logistics; cf. Figure 2.14), which can involve internal and external residue generators, waste disposal companies, dealers in used and secondary materials, specialized recycling companies, and scrap and recycling yards (Vaterrodt 1995, pp. 44).

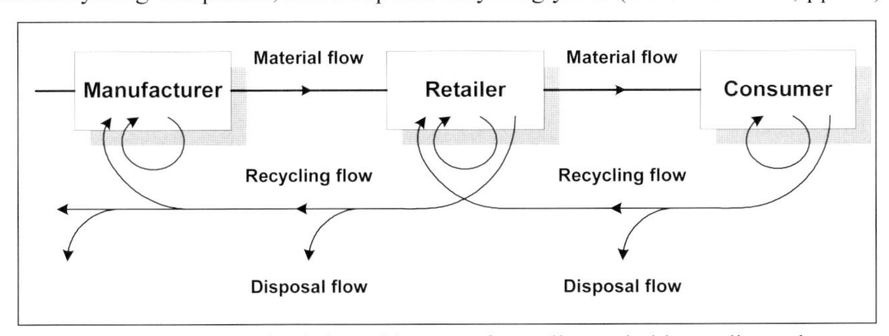

Figure 2.14: Supply chain making use of recycling and ultimate disposal

Product recycling exploits the fact that individual components of a product have a longer life than the product itself. Recycled products can be used as complete units or dismantled into individual components. The reusable parts are either used again for the same function as before in a new product (after any maintenance work needed) or technically modified before reuse.

Concerns that are relevant with respect to environmental goals in SCM are (Saunders 1997, pp. 174)

- the recovery, recycling, and reusing of materials and waste products,
- the safe disposal of waste products that cannot be recycled,
- supplier selection policies to support firms that conform to environmental standards with regard to air, water, and noise pollution,
- supplier and product selection policies that reflect concern for conservation and renewal of resources,
- safe testing procedures for materials and products, and
- concern for noise, spray, dirt, and vibration in operating transport facilities.

From the environmental viewpoint, the goal is to establish a "green supply chain". Practical guidelines have been established by the US Environmental Protection Agency (2000a), which has also described the effects that designing environmental managerial accounting systems has had on SC performance in three case studies (Environmental Protection Agency 2000b).

2.7.1 Recycling-Oriented Product Development

When disposal and recycling are borne in mind throughout product development, this can save costs and provide competitive advantages. For example, the voluntary acceptance of old PCs for recycling or disposal is a value-added service that improves customer relationships. Ethical considerations, legal regulations, governing protection of the environment, and product liability issues are reasons for companies' close attention to ecological matters.

Because of technical degrees of freedom that often exist in product design, engineering activities (cf. section 2.1) are particularly important in the product development process. The selection of materials and of operations for component manufacturing and assembly during design and work planning determines not only the manufacturing costs but also the recycling opportunities. Regulation 2243 "Recycling-oriented design of technical products" from the "Verein Deutscher Ingenieure" (VDI; Association of German Engineers) gives some assistance toward the achievement of recycling- and disposal-friendly designs. The main requirements formulated in it are the following:

- Basic material compatibility: The materials used in the production process should be compatible with each other in order to permit recycling together. Compatibility matrices can represent this property.
- Design for dismantling: This requires the use of fixings that are easy to undo (e.g., snap joints, clamping or twist locking rather than gluing) and easy access to the components.

Other measures are the identification of spare parts, restriction of the volume of materials to be dumped, and a corrosion-resistant product design. In addition, direct handling of the products should be aimed at, so as to avoid the need for packing material and packaging means.

A general issue is whether recycled parts are of an equal quality to new materials and whether any quality differences are relevant for their designated use. If necessary, the recycled parts have to be listed separately in bills of materials, in specifications, and in balance and mixing conditions of optimization models.

2.7.2 IS for Recycling and Disposal

Computer-supported aids are available for a recycling-oriented design: On the one hand, many companies use conventional software systems (e.g., CAx systems, ERP systems with materials databases), and, on the other, tools such as LASeR (**L**ife-cycle **A**ssembly **Se**rviceability and **R**ecycling Prototype, developed in the Manufacturing Modeling Lab at Stanford University) want to support an assembly-, service-, and disassembly-friendly product development [http://www-mml.stanford.edu/Research/Software/laser.html].

The Global Recycling Network [http://grn.com/] offers comprehensive information on the industrial use of recycling and recycled products on its Internet portal.

Environment information systems

Interplant environment information systems (cf. Bullinger et al. 1998) can make a major contribution to environment-oriented planning and control. Based on a suggestion from the German Office for the Environment, the central tasks of such systems are divided into the areas of material balancing, effects balancing, assessment, and associated reports. They do not contain only information relating to recycling and disposal, but also comprehensive data on law, regulations, emission-reduction measures, environmental statistics, garbage exchanges, etc. The use of an environment data catalog and the associated thesaurus developed under the leadership of the provincial government of Lower Saxony in Germany simplifies the structuring of the associated meta-data.

Recycling-oriented ERP systems

A precondition for the modeling of recycling processes in ERP systems is information on the composition and dismantling of products. This includes (cf. Kurbel et al. 1996, p. 56):

- Parts master data that describe the properties of parts (e.g., ingredients, dimensions, weight)
- Recycling graphs that show what components are connected with what others and how

- Recycling product structures that, by analogy with manufacturing BOM, specify the parts and components into which the products can be dismantled
- Disassembly plans that describe the associated operations.

Figure 2.15 illustrates the relationship between MRP and recycling data.

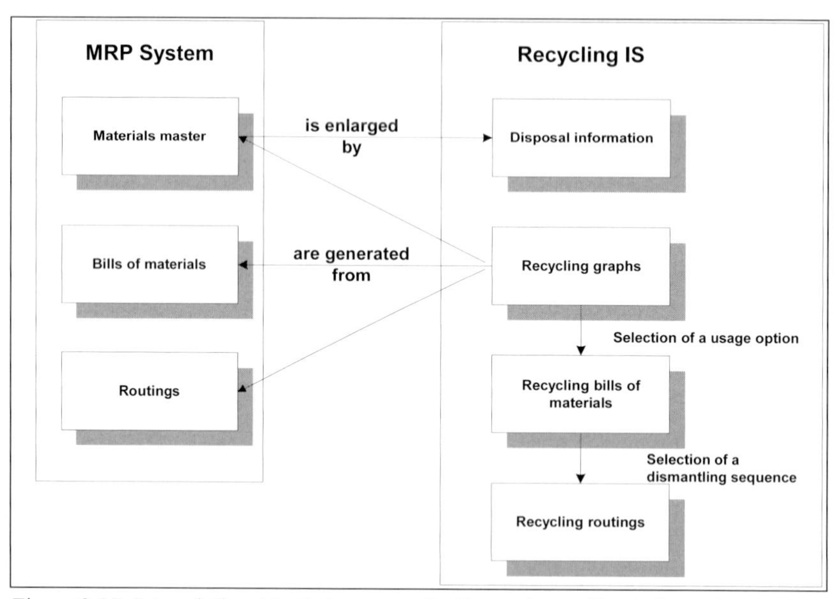

Figure 2.15: Interrelationships between production and recycling information systems (Kurbel et al. 1996, p. 57)

The functionality of conventional MRP systems is not sufficient to take account of such information. The extension for recycling-relevant aspects results in a procedure called integrated Production and Recycling Planning and Control (PRPC) (Rautenstrauch 1997). These PRPC systems could, for example, take account of the fact that the direct return of secondary components into the production process lowers the gross demands and may also shorten the lead-times.

The use of PRPC systems requires many changes relative to conventional production planning and control (cf. Kurbel/Rautenstrauch 1997):

- *Basic data:* To supplement the master data for the MRP systems, additional basic data (e.g., dismantling routings) are needed in PRPC systems.
- *Quantity planning:* This procedure must take into account those quantities that come back directly into the production process as recycled goods ("immediate recycling"). For these goods inventories emerge; attention must be paid to potential quality differences to the original materials.
- *Scheduling:* Extensions to existing scheduling procedures are necessary, because cyclical material flows and time-related dependencies can arise between production and recycling operations. For example, priority rules should ensure that recycled goods that cannot be stored (or stored only for a limited

time) are used as soon as possible. Furthermore, the capacities for processing and disposal resources must be observed.

- *Consideration of uncertain events:* If the rejects required for recycling arise at different intervals or in variable quantities, the determination of lead-times and capacity requirements becomes more difficult. It may prove necessary to determine probability distributions for the supply of the recycled goods and to take these into account in the planning process. However, this contradicts the usually deterministically operating MRP systems.

Dismantling Systems

Dismantling decisions affect, for example, the question as to whether specific recycled goods can be re-used for the repair of equipment or whether they must be scrapped. An example of support for such decisions by IS is DISRINS (DISmantling and Recycling INformation System (Scheuerer 1995). This system considers the reuse capabilities of components and the disposal costs in determining proposals to the planner. Possible benefits of dismantling systems lie, in particular, in the fast determination of economical dismantling steps, in preparing reasonable dismantling decisions, in providing information about joints that are critical in dismantling, and in savings achieved by the re-use of recycled goods (Mertens 2000, p. 247).

The International Dismantling Information System (IDIS) is a PC-based system which supports the identification of component materials in end of life vehicles [http://www.idis2.com/intro_eng.html]; twenty car manufacturers cooperate in this activity. Two modules have been developed: IDIS Office is offered for vehicle manufacturers and IDIS Plant for scrap dealers. The database contains data for more than 350 car types and about 20,000 parts.

Information Systems for Reverse Logistics

IS for reverse logistics differ from those for new products. Reverse logistics are often an exception-driven process; the data for the items entering the recovery chain are frequently of poor quality, the supply of used or remanufactured parts is sometimes regionally bounded, their promotion is often customer driven, and it may be difficult to identify the parts needed (Kokkinaki et al. 2000).

SAP provides functionalities that take environmental and recycling-oriented considerations into account in several R/3 modules:

1. *Recycling Administration (REA) function group: This provides the R/3 user with the capability to perform the settlement in accordance with the German Recycling and Waste Disposal Law and with EU Regulation 64/92. It uses existing master data and transaction data from the MM and SD modules to determine the billable packaging. The report system simplifies the cost control.*

2. *mySAP Environmental, Health & Safety (EH&S) provides the following capabilities:*
   - *Product safety*
   - *Dangerous goods management*
   - *Industrial hygiene and safety*
   - *Occupational health*
   - *Waste management.*

   These applications are integrated with mySAP solutions, especially in materials management, manufacturing and distribution logistics, plant maintenance, cost accounting, and HR management. Tasks supported are the
   - *generation of material safety data sheets to accompany the shipment of chemical products,*
   - *provision of substance information and substance reports at the inquiry and quotation stage and throughout the distribution logistics phase,*
   - *automatic check of the suitability of all means of transport for shipping dangerous goods,*
   - *substance notification data (legal registration) in a recipient country, and*
   - *waste disposal management and precise allocation of associated cost.*

   The substance database at the heart of EH&S manages information about properties and compositions of pure substances, preparations, and mixtures. This database also serves as a basis for environmental IS, product quantity restrictions, classifications of dangerous goods, and waste descriptions. Interfaces simplify the transfer of data from existing applications and allow importing of official data, such as from the EU list of dangerous substances. In addition to the substances and product database with predefined and user-defined properties, the following are also available:

   - *A phrase catalog with translations for use in multinational companies*
   - *Mechanisms by which the substances database can be distributed by way of various systems (e.g., from the parent company of a group to its subsidiaries, or in an SC)*
   - *Reports (e.g., safety datasheet, accident instruction sheet).*
3. *Functions in the MM and PP modules: The material BOM is a central data object in the R/3 System, which can be accessed company-wide: If important properties for recycling are entered in the material master data these can be used by all functional areas.*
4. *Industry solutions for the process and automotive industries: R/3 provides functionality for production planning and control in the process industry with PP-PI. This module uses some functionality of the EH&S component. In addition to the industry-oriented service characteristics of the SAP Automotive solution, the system also contains a function to support spare parts traders; this permits, for example, systematic recycling of selected returned parts.*

A special task within the avoidance of untimely final disposal is the redeployment of used machinery. Some ERP systems include asset management modules. However, these systems are usually in-house oriented and do not support the re-use of assets in other plants belonging to the group or in other companies. Asset Redeployment Management Systems (ARMS) are web-based applications that announce the availability of a used asset first within the group and, once an item-specific time interval has passed, to a global B2B exchange (cf. Figure 2.16).

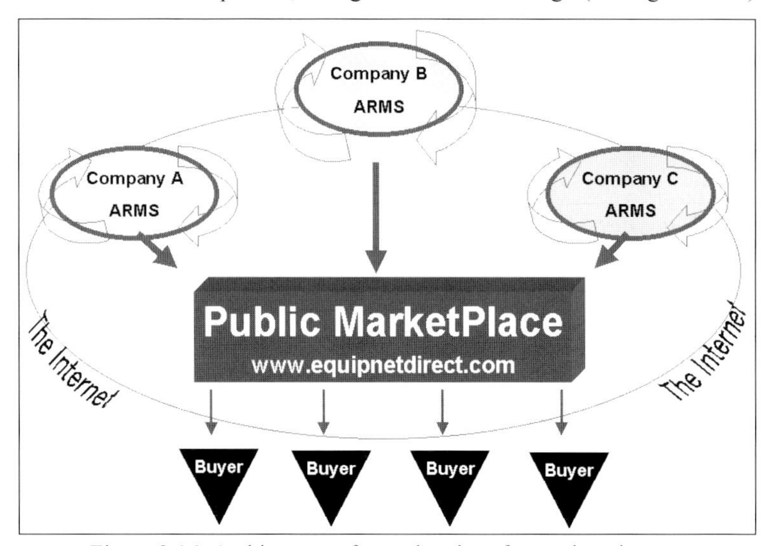

Figure 2.16: Architecture of a marketplace for used equipment

These systems may be implemented as add-ons to asset management modules of ERP systems, providing additional attributes about the items that may be relevant in a selling activity. New intermediaries are establishing Internet marketplaces for specific industries. These markets may be complemented by electronic auctions.

EquipNet Direct [http://www.equipnetdirect.com/] is offering redeployment services. It provides a turnkey solution for asset recovery by combining Internet trading communities with traditional asset management services. The auction place allows sellers to capture the market price on the product, often resulting in significantly higher asset recovery values. The services generate a higher return on assets for sellers as well as significant savings for buyers, and may also reduce the disposal of machinery for which a local dealer cannot find a customer.

EquipNet Direct maintains partnerships with industry providers of traditional asset management services that bring expertise into the areas of equipment appraisal, plant liquidation, consignment, outright purchase, refurbishment, start-up, training, re-installation, line integration, leasing, removal, preparation for shipment (crates, containers), and transportation. Thus, the Internet may increase the transparency in the used-asset market but does not provide all the technical expertise needed to deal with the assets.

3 SAP's Supply Chain Management System

Having discussed the principles of SCM in the previous chapters, we now consider the realization of this concept in SAP environments. We first give an overview of SAP's SCM Initiative and describe its main components

- Advanced Planner and Optimizer (APO)
- Business-to-Business Procurement, and
- Logistics Execution System (LES)

in sections 3.1 to 3.3. In section 3.4 we discuss possible architectures for the interactions between the IS from a hierarchical planning perspective. The text is based on SAP sources [e.g., from http://www.sap.com/solutions/scm/]; additional information can be found, for example, in Bartsch/Bickenbach (2001), Bartsch/Teufel (2000), Eddigehausen (2000), Grünewald (2000), Meyer et al. (2000), and Schneider/Grünewald (2000).

Recently, SAP has been presenting its SCM solutions under the header of mySAP Supply Chain Management, and it also includes the following components under this blanket:

- Business Information Warehouse (SAP BW)
- Materials Management (SAP MM)
- Production Planning (SAP PP)
- Sales and Distribution (SAP SD).

3.1 Advanced Planner and Optimizer

APO provides functions for intra- and inter-company planning of SC and for scheduling and monitoring the associated processes. The APO package consists of several modules that use a shared database (cf. also Figure 3.1):

1. *Supply Chain Cockpit (SCC):* A graphical "instrument panel" for SCM.
2. *Demand Planning (DP):* Statistical forecasting techniques and other methods for demand planning.
3. *Supply Network Planning (SNP):* Planning methods that consider the entire supply network.
4. *Production Planning (PP):* Techniques for short-term planning of material and production with due consideration for capacity constraints (finite planning).
5. *Detailed Scheduling (DS):* Methods for assigning resources and for sequencing production orders.
6. *Transportation Planning / Vehicle Scheduling (TP/VS):* Methods supporting truck loading, load consolidation, and carrier selection based on a bucketed shipment plan as well as on individual orders. Methods for scheduling vehicles and determining routes with respect to different types of constraints.
7. *Available-to-Promise (Global ATP):* Multi-level, rule-based availability checking with due consideration for inventories, allocations, production and transportation capacities, and costs.

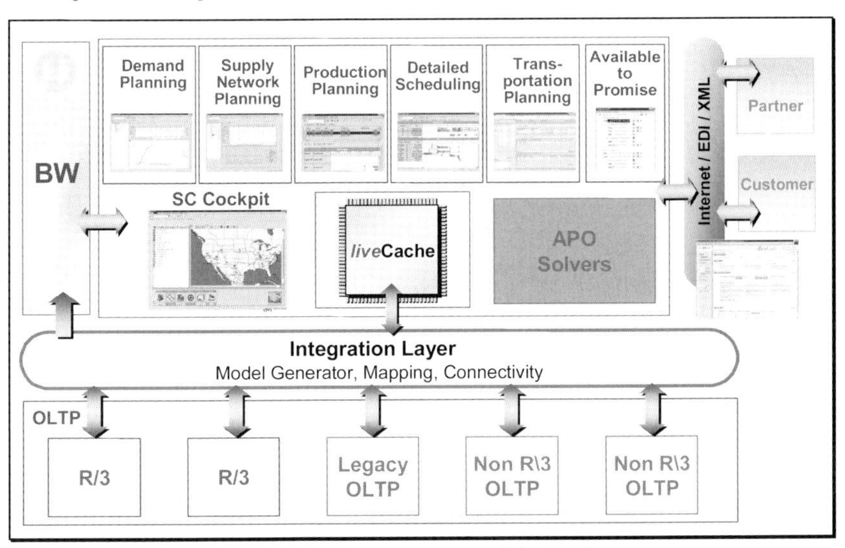

Figure 3.1: Architecture of the APO

Whereas we discuss the functionalities listed under 1, 2, 6, and 7 successively in separate sections, components 3 to 5 are grouped under the term "Advanced Planning and Scheduling" (APS), which is sometimes also used by SAP.

3.1.1 Supply Chain Cockpit

The SCC provides advanced visualization capabilities for planning and controlling both intra- and interorganizational logistical networks (cf. Figure 3.2). The SCC covers the following components:

1. The Network Design permits modeling of a logistics network. It is possible to evaluate different scenarios and to apply optimization algorithms.
2. The navigation component of the SCC makes it possible to look at many features of this network.
3. The Monitor informs department workers and managers of unusual states: If predefined exception conditions occur, the planner receives problem-related messages (Management by Exception). According to how serious the problems are, SAP distinguishes between informative, warning, and error messages.

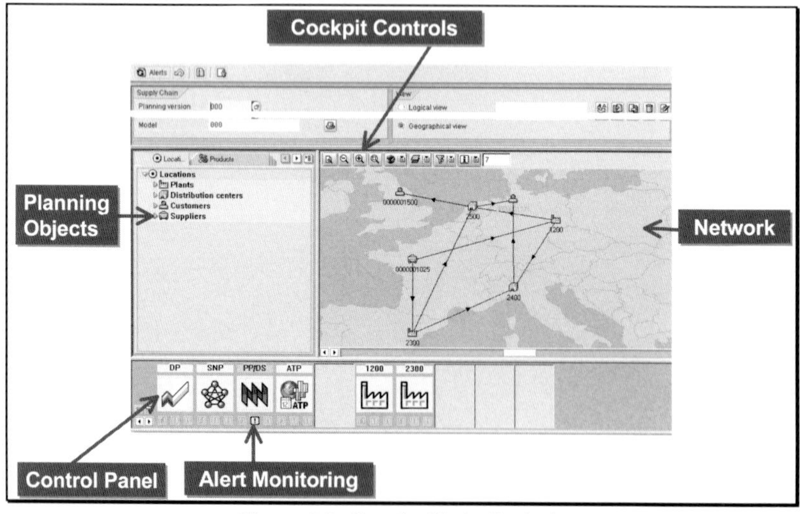

Figure 3.2: Supply Chain Cockpit

Typical users of the SCC are:

1. Network planners who support the establishment of a new supply network or the modification of an existing one.
2. Demand planners who need to access demand forecasts, receive messages from the Alert Monitor, and react appropriately.
3. Sales planners who create and change plans based on details supplied for the individual nodes.
4. Controllers who analyze the elementary data and, in particular, key performance indicators and investigate ways of improving the logistical structures and processes.

A network is based on nodes (locations) and arcs (transportation lanes) (Figure 3.3). A location is described by its type (e.g., supplier, factory, distribution center, and customer), its name, geographical information, details about resources and their capacities, a plant calendar, "profiles", etc. The profiles can contain various model variants with different degrees of accuracy. Arcs mark the routes that physical goods may take between the locations. Network models can be developed or modified by drag-and-drop procedures in the Supply Chain Engineer.

Whereas the network versions on a high aggregation level are used for planning purposes and simulations, more detailed versions are applied for scheduling and control. The products must be assigned to the locations with appropriate priorities.

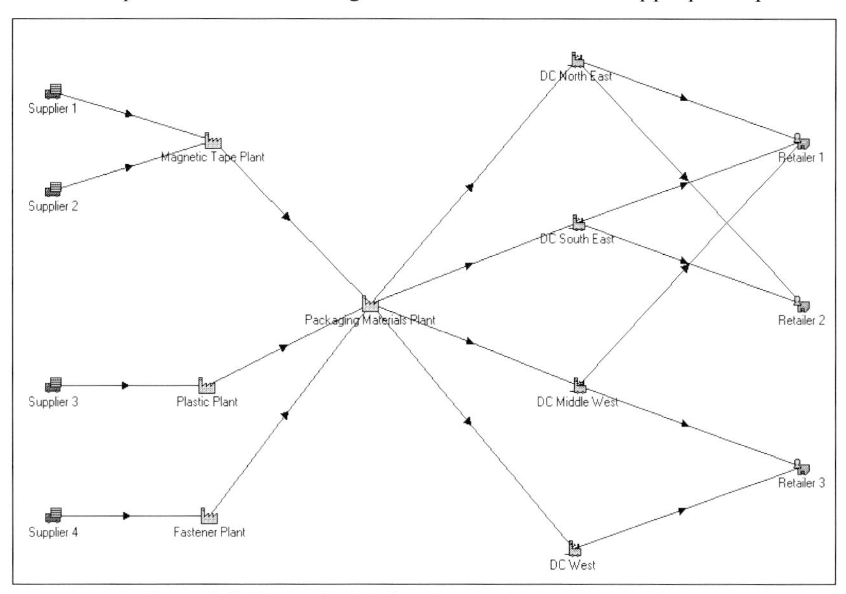

Figure 3.3: Network with locations and transportation lanes

The Transport Planner can assign attributes, such as a transportation method, resources, priorities, calendars, durations, and validity periods to the network. Quotas with which individual locations and transport lanes are involved in a logistics process, lot size profiles, etc. can be specified for the disposition heuristically. The lot size profiles describe the rules to be observed in determining lots, e.g., with regard to rounding rules.

The SCC permits typical queries, e.g., about customer orders, warehouse inventory levels, restocking, current capacity situations, production and transport quantities, but also about such key figures as the ability to deliver, shortfall quantities, deviations from the delivery schedule, development duration, and market share.

User query profiles are saved for executing standard queries. To simplify navigation, one can zoom subareas of the net or filter out information.

The Business Information Warehouse (BW) function is designed primarily for management information; "Information Cubes" (InfoCubes) are typically used in this component. Figure 3.4 illustrates an InfoCube with the dimensions material, region, and period.

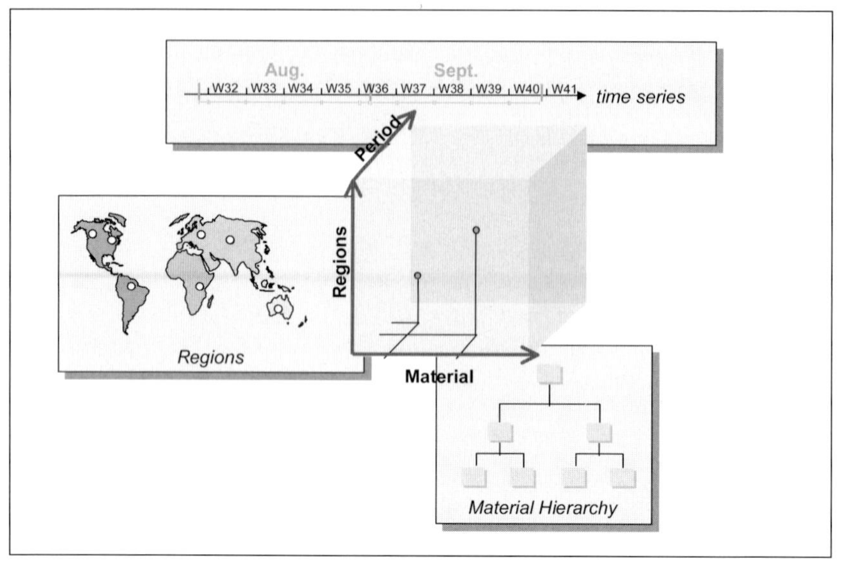

Figure 3.4: Visualization of an InfoCube

When they apply the Alert Monitor, the department staff and managers concerned are informed in real time about any exception situation defined as relevant. Both an individual exception status and aggregated data can be presented. Connections exist, e.g., with production and distribution planning. The individual messages are assigned according to priorities. A drill-down function makes it possible to investigate the cause of a warning in detail and to show any connections discovered. SAP refers to this content as the "Problem Resolution Screen". Examples for exception messages are:

- Invalid overlapping of operations
- Violation of time constraints
- Excessive or inadequate inventory
- Capacity overloading of resources.

Special functions in the SCC simplify updating of the master data. For example, one can click to accept locations from stored maps or assign specific parameters not only to a single entity but also to a group of elements (e.g., a product group); this procedure is called "mass updating".

3.1.2 Demand Planning

An important precondition for preparation of a realistic production plan is the ability to forecast demand with adequate precision. Inaccurate forecasts are result from inadequate information exchange between different participants in the SC (cf. the discussion of the bullwhip effect in section 1.2.1; Fisher et al. 1994). Two simple examples can be cited to illustrate the advantages of information exchange:

1. A retailer orders an article with seasonal demand, depending on school holidays, as follows: 5,000 items each for the southern region of a country in March, for the central region in April, and for the northern region in May.

   As a result of its own purchasing and manufacturing situation, the manufacturer knows in March that it can supply this retailer with only 10,000 items during the entire season.

   Typically, the retailer does not become aware of the impossibility of getting supplies until May, when it tries to place the last order. However, shortly before this, the retailer may have noticed low demand in the South and started a regional sales promotion offering the article at a reduced price. Had the retailer known about the limited delivery capability of the manufacturer, the quantity available could have been distributed differently to the regions and the price reduction avoided.
2. A retailer plans a sales promotion for a specific product. Based on experience from an earlier promotion, the retailer estimates that the sales volume will increase to 50 percent above the regular level. The retailer shares this forecast with the supplier. The manufacturer then forecasts that, because of this promotion, additional demand will also arise in neighboring stores. It recommends to other customers that they adjust their sales forecasts and the resulting warehousing plans.

The "Demand Planning" (DP) component of the APO is used to forecast the products sold. This APO component generates the sales plan of a business area, which serves as the starting point for the subsequent planning activities.

Basis of demand planning

A sales plan is frequently based on historical data, which may be available directly from ERP systems or archiving systems or, in aggregated form, from a data warehouse. The connection of sales planning with the SAP Business Information Warehouse gives companies the capability to evaluate basic data from ERP systems with respect to specific multi-dimensional criteria (e.g., product groups, regions, and time intervals). The analysis returns information about the factors that affect the demand plans. Rules must be defined for extraction and aggregation of the ERP data. They specify how and when the derived data have to be corrected as a result of changes in the elementary data. The time dimension for data is of major importance in BW systems (Stock 2001). Coordination needs result if different partners in the SC use different calendars (e.g., plant calendars, location-dependent warehouse calendars).

Figure 3.5 illustrates the interaction between DP and the BW. Online Analytical Processing (OLAP) servers provide the end-users with powerful aggregation and retrieval functions for data analysis.

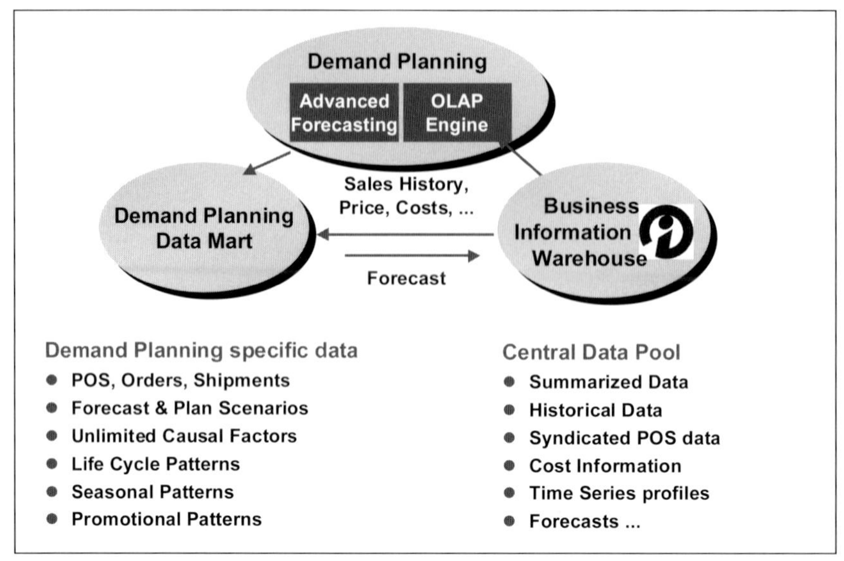

Figure 3.5: Datasets in DP and BW and data flows between DP and BW

Because historical data are of only limited use for forecasts, additional knowledge should also be used as input for forecasting. For example, SAP R/3 permits the use of data from external market research institutes [e.g., from Nielsen or Information Resources, Inc. (IRI)].

Forecasting methods

Many forecasting methods have been developed in the literature; some of them have been implemented and are offered in method bases (cf. Mertens 1994a; Brander 1995; Makridakis et al. 1997). APO provides the most important methods for practical use:

- Smoothed average values
- Exponential first- and second-order smoothing
- Trend models
- Trend-season models (Holt-Winters model)
- Croston method for sporadic demands (Croston 1972)
- Multiple linear regression analysis.

The user can configure the different methods individually; DP can automatically select a suitable model variant for time series models. The planner may specify, for instance, how untypical values are to be handled or how the number of work days per period is to be considered.

APO provides more comprehensive forecasting capabilities than those available in the R/3 SD module. In particular, APO contains the following extensions to R/3:

- Correlation and regression methods for the analysis of historical data
- Combinations of forecasting results ("Composite Forecasting")
- Option of including causal factors (e.g., advertising actions, marketing budgets)
- Consideration of sporadic demand.

User-defined planning views and interactive planning books allow different departments of a division or several plants of an SC to cooperate in the forecasting process. Data from several organizational units or enterprises are stored in a common database. APO offers a function-based planning feature for companies that need to forecast the demand for products with a wide range of variants.

Collaborative Planning, Forecasting, and Replenishment

Collaborative Planning, Forecasting, and Replenishment (CPFR) is the term for a business model involving several companies in an SC: It starts with joint agreements on business practices and conditions and ends with largely automated stocking of their warehouses. SAP uses the "CPFR Voluntary Guidelines" prepared by the CPFR Committee [http://www.cpfr.org/] as its principal reference source in this area. Representatives of approximately 70 manufacturing and trading companies are active in this group; the committee was set up by the VICS (Voluntary Interindustry Commerce Standards) Association [http://www.vics.org/].

CPFR is directed at improving the inter-organizational partnership between suppliers and customers through jointly administered information and cooperatively managed processes to achieve win-win situations. The CPFR model contains a comprehensive list of terms that must be added to a repository, which can be accessed by all participants in the SC. In particular, CPFR is intended to attenuate the bullwhip effect. It is possible to save money if the partners in the SC improve coordination and, for example, avoid multiple administration and repeated use of planning methods and databases. The costs of data exchange between participating companies have fallen with recent advances in telecommunications, especially the Internet, and have thus declined in importance.

According to a 1996 study in the US (cf. CPFR 1998, p. 7), about 8 percent of the desired purchases could not be satisfied because of inability to deliver. This caused a sales value loss of more than 3 percent; it should be noted that in about half the cases the customers ordered substitute articles, the majority of which were manufactured by competitors. Material supplied by the CPFR Committee highlights the following weak spots in restocking within supply networks:

- Most companies generate multiple, interdependent demand forecasts for different purposes.
- Most forecasting is done at an aggregate level of detail which focuses on a product category or family, a market or region, and a period of weeks or months.

- Forecast accuracy is not measured regularly.
- Operational forecasts usually focus exclusively on interaction between only two nodes in the value chain. The forecasts do not always use identical time intervals throughout the SC.
- Manufacturing usually pushes inventory to its distribution centers on the basis of production economics and not of customer demand (pull principle).

Guidelines for CPFR are:

- The partners develop an outline plan in which core process activities are assigned to the participants.
- The cooperating companies jointly prepare a forecast of the consumers' demands. This forecast serves as the basis for the plans and schedules to be agreed on ("Shared Forecast"). Because of the different requirements of the participating companies, the prediction should be rather detailed.
- Together the partners attempt to avoid inappropriate schedules in the material flow. For example, small order quantities from retailers may result in small utilization degrees of the production capacities or in excess capacity for satisfying short-term delivery requests. When the retailer orders larger sizes, a better coordination with the production plan of its supplier may result. In another case, the willingness of a retailer to avoid requesting short delivery times can have the result that the manufacturer changes from "make-to-stock" to "make-to-order" production, thus reducing lockup of capital and uncertainties for scheduling.

Figure 3.6 shows that the CPFR process model is divided into three levels (planning, forecasting, and replenishment) and comprises nine steps.

Step 1: Develop front-end agreement

Manufacturers, retailers, and distributors develop rules for the cooperation. The partners announce their expectations and define the necessary resources. These contracts define the joint use of information, the rights and liabilities of the partners (e.g., confidentiality, provision of resources, distribution of profits), and the criteria and metrics used to measure the effectiveness and success of the CPFR process (e.g., forecasting accuracy, profitability of the investments, turn-over of inventory, maximum delivery time, reliability in meeting deadlines, process costs).

Step 2: Create joint business plan

In this step, the manufacturer and retailer exchange information about their company strategies and develop a joint strategy. This results in certain agreements, e.g., on minimum ordering quantities, multiplier factors (with particular reference to packaging units), lead-times, and ordering intervals. The goal of this step is to minimize the number of exceptions. This phase also includes agreements about marketing and sales promotions, e.g., for advertising campaigns or temporal price reductions. Thus, the decisions made in Step 2 form an important basis for joint forecasts.

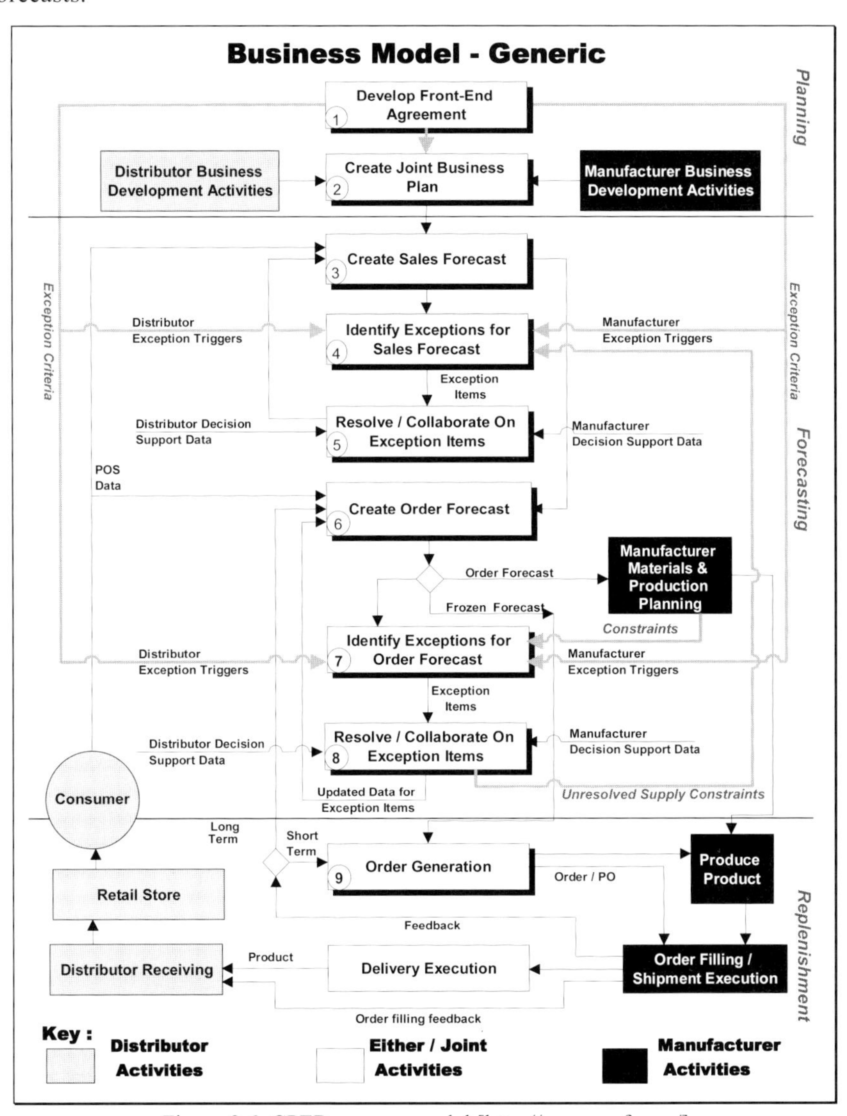

Figure 3.6: CPFR process model [http://www.cpfr.org/]

Step 3: Create sales forecast

This forecast is produced on the basis of the point-of-sale (POS) data and from information on special factors and planned sales promotions (cf. Step 2).

Table 3.1 shows the four different scenarios considered in the CPFR guidelines; whereas in scenarios A, B, and C activities are controlled by the retailer or distributor, it is the manufacturer that controls them in scenario D. Figure 3.7 provides an impression of the individual measures to be realized in Step 3 (creating the sales forecast) for scenarios A, B, and C.

Table 3.1: Four scenarios for cooperation between the SC partners

Scenario / Activity Lead	Sales Forecast	Order Forecast	Order Generation
Scenario A	Retailer / Distributor	Retailer / Distributor	Retailer / Distributor
Scenario B	Retailer / Distributor	Manufacturer	Manufacturer
Scenario C	Retailer / Distributor	Retailer / Distributor	Manufacturer
Scenario D	Manufacturer	Manufacturer	Manufacturer

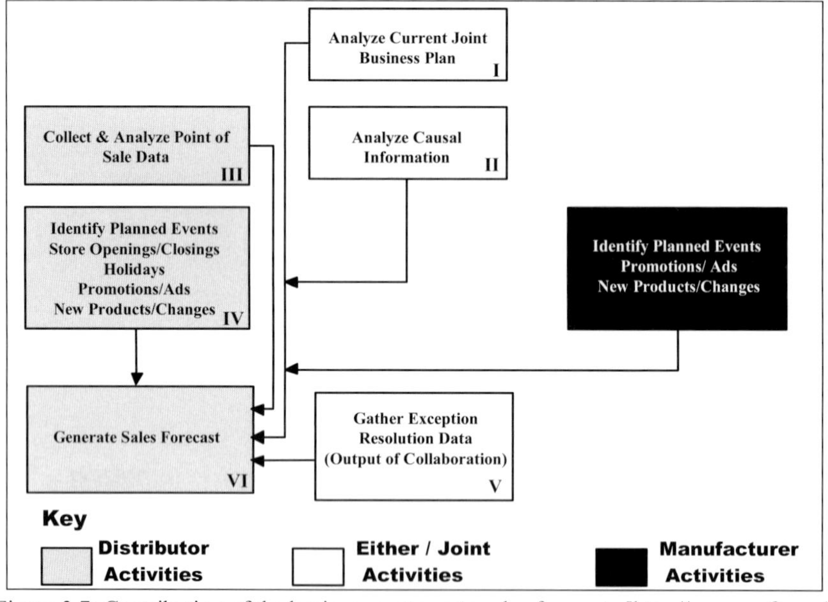

Figure 3.7: Contribution of the business partners to sales forecasts [http://www.cpfr.org/]

The CPFR guidelines do not give a detailed description of the methods to be used for a joint forecast. The following alternatives are examples that could be considered, either singly or in combination:

- Develop a forecasting system that processes time series data from various sources (e.g., POS data from retailers, outward stock movements from distribution centers, orders received by the manufacturer). For example, a multiple regression model is conceivable. The system itself can use regression analysis to determine the factors that weight the impacting quantities.

- Various participants in the SC use their "in-house" systems to prepare individual forecasts, which are integrated to give the joint forecast. The following variants can be envisaged:
  a) Simple arithmetic average of the individual forecast values
  b) Weighted arithmetic average of these values:
    b1) Weighting according to the respective sales volumes of the companies.
    b2) Weighting according to the demand quantity that the forecasting company provided in the supply network; orders placed with companies outside the CPFR system are disregarded.
    b3) Weighting according to the forecasting quality in the past. For example, a method based on an algorithm published by Chow (1965) can be used. In this case, three forecasts are calculated using the method of exponential smoothing with different smoothing factors α_i (e.g., $0.1 \leq \alpha_i \leq 0.3$). Assume that α_1=0.15, α_2=0.20, and α_3=0.25 are set. It is possible to determine later which α_i gave the best forecast; if this is α_2, the previous α-values are used again in the next period.
    If α_1 yielded the best forecast, α_1=0.10, α_2=0.15, and α_3=0.20 are used for the next prediction. If the best forecast was obtained when α_3=0.25 was applied, the new parameter set is {0.20 0.25 0.30}. Depending on the type and quality of the forecast, one can choose different deltas by which the α_i are adjusted (0.05 in the example).
  c) Other possible ways of deriving the joint forecast from the forecasts of the individual partners are:
    c1) Choose the most pessimistic forecast.
    c2) Choose the most optimistic forecast.
    c3) When the supplier's forecast differs by more than, say, 30 percent from that of the customer, the supplier uses its own forecast, and in all other cases, that of the customer. In the first case, the customer is informed of the discrepancies in the planning assumptions.

  Because different methods can be desirable in various industries and supplier networks, SAP does not prescribe the algorithms to be chosen; rather, APO provides a macro language which the SC partners may apply to build their own joint-forecasting system.
- The partners also agree on a tolerance corridor. Should the deviation of actual from planned data fall outside this corridor, a personal meeting or an electronic conference could be arranged with the aim of reaching an agreement. Such conferences might be supported by IT systems, for example by using the principles of computer-supported Delphi forecasts (Brockhoff 1979) which allows all the involved parties to coordinate their individual forecasts in a well-defined way. The mean value of the forecast and the deviations of the individual partners' own forecasts from it and the scatter can be calculated and the participants are informed of the results via E-mail. The planners can revise

their forecasts in the next step. This procedure is repeated until a convergence criterion, which can be customized by the SC partners, is reached.

Step 4: Identify exceptions for sales forecast

This step determines those forecast objects for which the actual requirements differ from the forecast by more than a specified tolerance threshold (as customized in Step 1) are determined. The causes do not lie only in incorrect forecasts, but also in external disturbances, such as a reduction in manufacturing capacity resulting from a strike or machine failure. The SAP Alert Monitor can be used to signal the exception situation to the members of staff concerned.

Step 5: Resolve/collaborate on exception items

Physical and electronic conferences are envisaged in this phase. However, shared databases are also used, which, for example, provide information on new events that exert strong influences on sales. This results in a modified forecast.

Step 6: Create order forecast

The demand forecasts and the inventory information (on physical inventory, open orders, inventory in transit) are now combined to predict what orders will be received. The agreements reached in Step 1, for example on measuring safety stock, order quantities or lead-times, must be taken into account. The result of this step is a time-differentiated forecast of order quantities. The participants in the SC should be able to rely on their partners' schedules not changing within a certain time period ("frozen zone").

Step 7: Identify exceptions for order forecast

In this step (which is analogous to Step 4), the order arrivals that are at odds with policies agreed on by manufacturers and retailers are specified. For example, there could be demand that cannot be met in the available time because of inadequate production capacities or logistical difficulties.

Step 8: Resolve/collaborate on exception items

Step 8 follows Step 7 in much the same way as Step 5 follows Step 4. Once again, information to support decisions must be obtained from the databases. Then it has to be determined whether the exceptions can be ignored or, if not, what action should be taken; it might, for example, be deemed necessary to purchasing from suppliers that do not belong to the SC. Information on the deviation may need to be returned to Steps 4 and 6, and these may have to be executed again.

Step 9: Order generation

In this step, the forecast orders are converted to fixed orders. Confirmations of orders must be sent to the customers.

APO covers parts of CPFR by its Collaboration Engine. Figure 3.8 shows some features of this engine. The Planning Books contain input data for the collaborative planning process such as the individual and the resulting joint forecasts and other data based on them. The Bid Invitations are addressed towards the suppliers that might offer certain products. The Collaboration Engine is responsible for planning and controlling the process of inter-company cooperation. It contains the application logic and uses already available components, such as InfoCubes, extended macros, and the Planning Book Designer. The communication layer mainly provides tools for data exchange with partners.

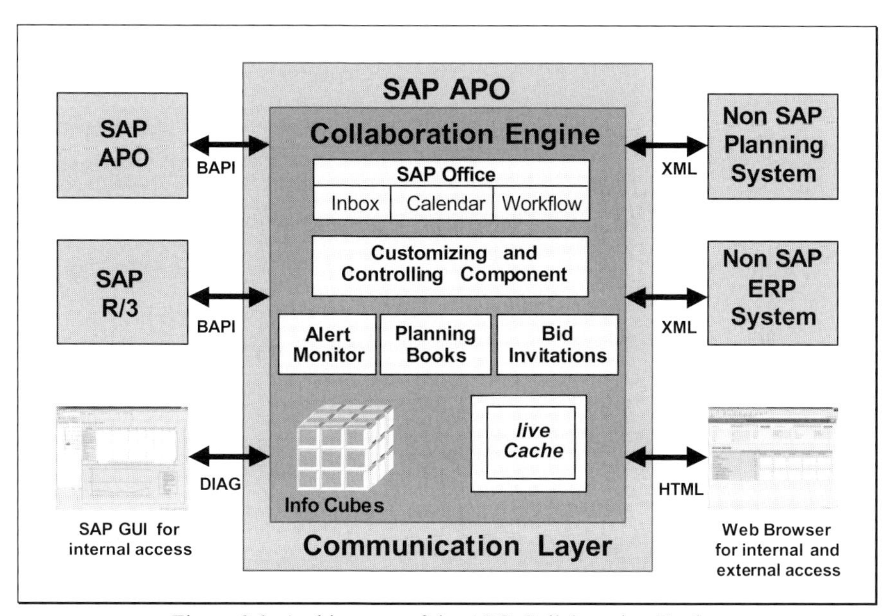

Figure 3.8: Architecture of the APO Collaboration Engine

3.1.3 Advanced Planning and Scheduling

Advanced Planning and Scheduling (APS; cf. section 2.4.2) systems contain functions that go far beyond those provided by ERP systems. PriceWaterhouseCoopers, without describing further details, rated the advantages of the use of various APS functions for different industries and process types (Table 3.2). A typical feature of APS systems is that different planning methods operating with different time horizons are provided (cf. Figure 3.9).

Table 3.2: Use of different APS functionality in various industries and plant types (PriceWaterhouseCoopers 1999, p. 108)

Applicability of APS functions	*(1)*	*(2)*	*(3)*	*(4)*	*(5)*	*(6)*	*(7)*	*(8)*
Network planning	L	L	H	H	M	M	M	M
Sales and operations	M	H	M	M	H	H	M	L
Demand planning and communication	L	M	M	M	M	H	H	H
Supply planning	M	M	M	H	H	M	L	L
Available/capable to promise	H	H	H	H	H	M	L	L
Distribution planning	L	M	M	L	L	H	H	M
Manufacturing planning and scheduling	H	H	L	M	M	L	L	L
Deployment planning	L	M	H	M	M	H	M	L
Warehouse management	L	M	M	L	L	H	H	M
Transportation planning and scheduling	L	M	L	L	L	H	H	M

Legend:

Plant types: (1) Batch manufacturing; (2) Electronics assembly;
(3) Pharmaceutical; (4) Process (continuous); (5) Process (batch);
(6) Consumer packaged goods; (7) Wholesaling and distribution; (8) Retailing.
Importance: (H) High; (M) Medium; (L) Low.

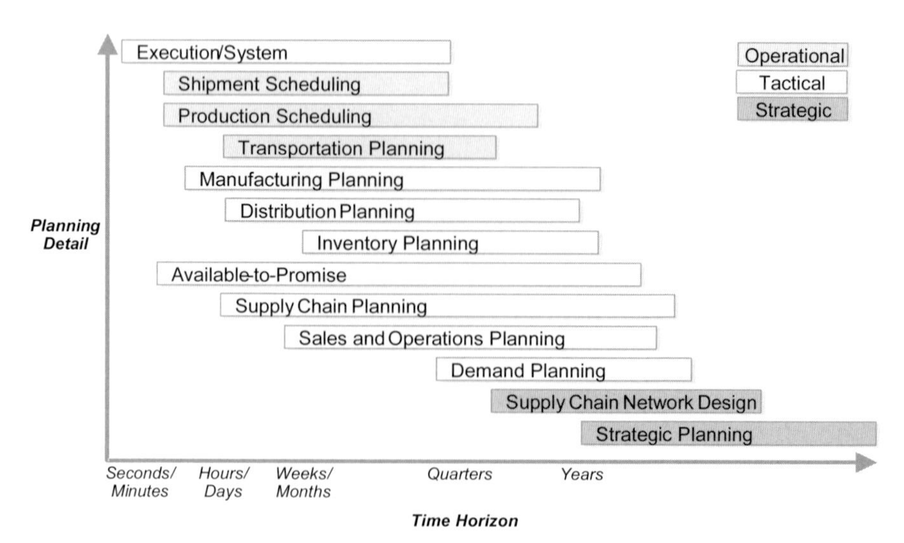

Figure 3.9: Planning horizons of different APS methods (Lapide 1998)

SAP tends not to use the term APS for its product offering; the functions that other vendors group under APS are provided in APO under the terms

- Supply Network Planning (SNP),
- Deployment, and
- Production Planning/Detailed Scheduling (PP/DS).

Figure 3.10 gives an overview of the sequence in which the APS methods are typically applied.

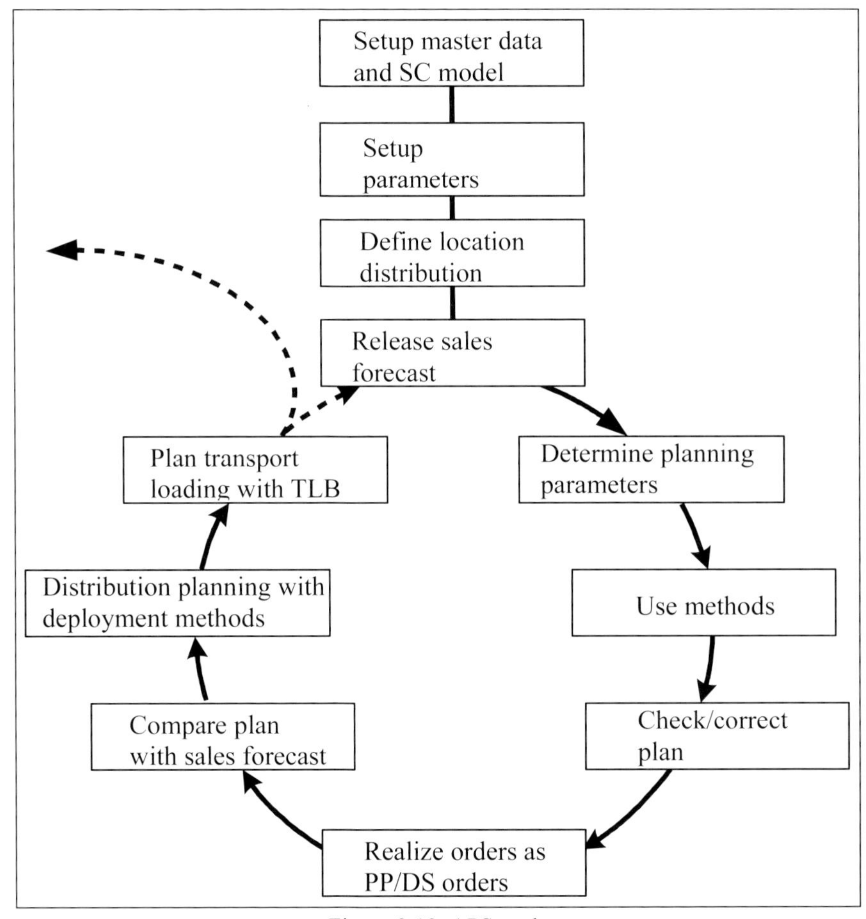

Figure 3.10: APS cycle

The implemented methods greatly exceed the capabilities that R/3 offers in the PP module. Therefore, some of the master data required by PP and the APS components are different. The APS functions include customization capabilities that can be used to specify the routinely applied planning procedures. In particular, APS has the following features that PP does not:

- Use of optimization procedures, heuristics or a simple dialog system
- Task-specific user interfaces
- Different aggregation levels and planning accuracy
- Extensive links ("pegging" between data objects [cf. Toomey 1996, pp. 70; Nicholas 1998, p. 811], e.g., between customer, production, and transport orders)
- Alert mechanisms to deal with exception situations.

Similar to ERP systems (Dittrich et al. 2000), the APS function also contains a large number of parameters that make special procedure models useful for APO implementation. Cap Gemini Ernst & Young offers APO Supply Chain Solution Centers to support the project team with pre-configured software, an advanced development environment with state-of-the-art infrastructure, technical support and work facilities, and a tool named RapidAPS which contains detailed plans, methodologies, models, and templates for all processes for accelerating implementation time [http://www.us.cgey.com/ind_serv/services/erp/sap_approach_supply.asp].

Some of the methods described as optimization procedures in SAP documents are termed heuristics in decision theory because they usually do not find the exact optimum of the mathematical model. In SAP terminology, all methods that explicitly use an objective function are designated as optimization procedures and most others as heuristics. SNP allows the application of mathematical optimization methods but PP/DS does not.

Supply Network Planning

This APO module can be used to determine production, procurement, and distribution schedules at the tactical level, namely with average accuracy; it concentrates on the associated quantities, additional capacities (e.g., with overtime), and the use of external suppliers. SNP is based on periods whose length is often a week or a day. For this purpose a so-called bucket profile may be defined; typical cases are

- Planning horizon 3 weeks, time-bucket one day
- Planning horizon 5 or 9 weeks, time-bucket one week
- Planning horizon 12 months, time-bucket one month.

SNP provides different priority classes based on external priorities, the maximum length of time by which the delivery date can be exceeded, and the different forms of the associated penalty costs. A distinction can be made between customer orders and internal orders. The four planning horizons, Deployment Push Horizon, Production Horizon, Stock Transfer Horizon, and Supply Horizon, differ in whether or not distribution, production, warehousing, and/or demand schedules can be changed.

SNP provides three different planning methods (cf. Table 3.3 and 3.4).

Table 3.3: Similarities and differences between SNP and PP/DS

Property	SNP Optimizer	SNP Heuristic	SNP CTM	PP/DS
Typical temporal accuracy	1 day			Minute
Typical planning horizon	12-18 months			4-8 weeks
Accuracy of modeling production resources	Limited bucket resources			Different capacities per interval; concentrating on selected resources (bottlenecks); single and multi-activity resources.
Temporal unavailability of resources	Determined by resource calendar			Shift changes, breaks
Limited handling, transport, and storage resources	yes	yes	no	yes
Shift planning	Shifts are not explicitly considered but determined by calendars			Detailed shift planning
Considering additional shifts to avoid bottlenecks	yes	no	no	no
Considering setup times	no			yes
Determining earliest and latest start of processing	no			yes
Considering restrictions when determining lot sizes	yes			yes
Determining and reviewing safety stocks	yes			yes

The member of staff responsible for the application of appropriate planning methods is faced with a cost/benefit problem, because there is usually a trade-off between solution quality and computing time; the following aspects have to be considered:

- The accuracy of the available data
- The importance of determining an exact optimum

- The robustness of the solutions computed with regard to model and data changes
- The times that the planners have to spend on training, model formulation, solution determination, and interpretation
- The necessary computing time and memory requirements of computers.

Table 3.4: Comparison of methods available in Supply Network Planning

Property	**Mathematical optimization**	**SNP-heuristic**	**Capable to Match**
Solution quality	+	–	0
Consideration of production capacities	+	Done locally in a second step	+
other capacities	+	Done locally in a second step	–
Speed	–	+	0
Consideration of priorities	Priority given to PPM with lowest costs	–	Particularly for locations, materials
Quotas	Optimized	Predetermined	Predetermined

Legend: + ... well suited; 0 ... average suitability; – ... less well suited

SNP is organized in the following phases:

1. A sales plan is developed in the R/3 SD module, in APO, or in a non-SAP system. Material and capacity constraints are disregarded. This plan is released to SNP, where it is used as input for further planning steps.
2. A mid-term production and distribution schedule is determined by SNP. The following planning procedures are available in SNP:

   2.1 Use of linear or mixed-integer programming

   2.1.1 based on (non-aggregated) master data of production planning. The use of integer variables allows, for example, the feasible values for batch sizes to be restricted so that loading equipment (e.g., pallets or containers) is always filled exactly.

   2.1.2 for aggregated models. The following cases are conceivable:
   - Aggregation of elementary time-buckets to longer intervals and a truncation of the planning horizon, i.e., the latest time point included in the plan
   - Aggregation of products in product groups
   - Aggregation of individual orders in order groups.

   Another simplification is possible if capacities that will definitely not become bottlenecks are not included in APO models because they are mathematically redundant.

In generation of mathematical programming models, up to now matrix generators such as OMNI [http://www.haverly.com/omni.htm] or modeling languages like AMPL have usually had to be used to define the optimization model [http://www.ampl.com/cm/cs/what/ampl/]. In APO, SAP provides generic optimization models that the planner can customize to specific needs, e.g., by using drag-and-drop procedures. This approach is less flexible than employing a modeling language but simplifies the model-building process and may thus, together with the visualization capabilities, contribute to the acceptance of Operations Research methods by planners and managers. Different user groups may choose distinct approaches for model formulation and usage.

In the generic optimization model, the following cost parameters can be defined:

- Production costs
- Costs for additional production capacity
- Procurement costs
- Transportation costs
- Costs for additional transport capacity
- Storage costs
- Costs for additional storage capacity
- Costs for handling capacity
- Penalty costs for safety stock
- Penalty costs for late delivery
- Penalty costs for non-delivery.

The planner can assign linear weighting factors to the cost parameters by moving buttons on a control panel. These data are used by the algorithm to find an optimal solution: The system generates a production and transportation plan, which takes the interdependences between the different types of costs and between the relevant constraints into account. The solution providing the lowest costs is proposed as the optimal solution. The planner may explore the sensitivity of the solution and study different planning scenarios by changing the weightings on the cost parameters; he may, e.g., determine to what extent higher storage costs will generate additional transportation orders.

The integration of user-specific optimization algorithms and heuristics is facilitated through the Optimization Extension Workbench. This tool makes it possible to combine APO procedures with third-party or in-house optimizers to meet specific industry or company needs.

First experiences show that quite large linear programming models are formulated and solved. For example, in the Kimberly-Clark group [http://www.kimberly-clark.com/] models with up to 2,500,000 variables and 600,000 constraints have been formulated.

2.2 Use of the SNP heuristic

The SAP term "SNP heuristic" is a little confusing, because CTM must also be classified as heuristic. Like the optimization procedure described above, the SNP heuristic adds all planned demands for a product at a certain location to determine the total demand for each time-bucket. It then determines the supply capabilities and the corresponding quantities using predefined quotas for each source of supply. Capacity restrictions and partial availability are disregarded in this planning step. Thus, the resulting solution is not necessarily feasible. The planner may subsequently perform an interactive capacity leveling for each location and thus determine a feasible plan in a second step.

2.3 Capable to Match (CTM)

Particularly in high-tech industries, CTM is a popular heuristic which operates either with bucket-oriented demands or individual customer orders. In search of a feasible solution, CTM considers order priorities and categorized sources of supply (e.g., stock on hand, production capacity, safety stocks). Handling capacities and warehouse and transport resources are not taken into account. If no production capacity is available the demand initially remains uncovered. With regard to a certain supplier, CTM can consider constraints for the quantities to be provided by each location and for each product. Furthermore, location priorities can be defined and orders can be generated at one or more locations to cover demand. CTM considers several rules that the planner may have formulated.

CTM proposes the first solution without taking costs into account. In particular, no replanning of previously assigned orders is considered within a planning run. The resulting solution is usually not optimal. However, the underlying model is more detailed than the model of the SNP heuristic and thus CTM often provides better solutions.

To simplify the resulting planning procedures, APO provides various "meta-heuristics" (cf. Voss et al. 1998) that can be used in individual solution steps on certain sections of an SNP procedure. These simplifications often drastically reduce the computing time compared with that needed to solve the original model.

Heuristics or computer dialogs can be used to adapt the schedules created with one of the SNP methods to allow for constraints that remained consciously unconsidered in an earlier planning step. The resulting schedule is then made available to the "Production Planning and Detailed Scheduling" (PP/DS) component. This procedure matches the production plan with the actual status of the production system and, after considering bottlenecks, tries to create a feasible schedule.

The detailed plan determined in this manner can be compared with the upper-level plan. Significant changes may require the planning process to be repeated with changed premises.

SNP includes an alert mechanism, namely a warning function that sends information to the inventory planners when certain events arise. The planner can process this message in a problem solution window. Certain types of alerts may be automatically passed to another member of staff. SNP should be seen as an advanced, constraint-oriented production and distribution planning system supporting order management rather than as a system focusing on the cross-company cooperation within the SC. The inclusion of distribution planning has extended the decision models formulated in the literature and made them more practical. Forthcoming releases can be expected to offer further developments in the direction of cross-company coordination.

The methods available in SNP access data that are released from the data cube created in DP to the liveCache, a memory-resident data store that may occupy as much as 32 GB on Unix systems.

Deployment and Transport Load Builder

With the combination of production planning and deployment, an attempt is made to find an appropriate use of manufacturing, distribution, and transport resources with respect to the constraints existing in the SC. After use of SNP, the so-called deployment methods test whether demand exceeds supply or vice versa. If the planned quantities (supply and demand) agree with the actual values, the result of the deployment confirms the higher level plan. However, if the available quantities differ markedly, the distribution plan is adapted using one of the following allocation rules (so-called fair-share methods):

- Rule A: The same percentage value of the current demand is assigned to each demand node.
- Rule B: At every demand node, the same percentage of the planned inventory is fulfilled.

If larger quantities are available than demanded, pull, push, and pull/push strategies can be used to allocate the excess stock to the individual warehouses. The deployment procedure may also initiate restocking orders.

The orders confirmed by the Transport Load Builder (TLB) component are passed to transport scheduling. This scheduling procedure considers, for example,

- different types of transport equipment, the products that can be transported with them, and the associated transport costs,
- minimum and maximum weights of the transport loads,
- anticipation horizons during which transports can be performed in advance in order to satisfy the minimum transport loads,

- the maximum number of transport pallets for each means of transport, and
- the maximum volume for each transport means.

Manual interventions are again possible for interactive assignment of any transport orders not considered by the heuristic.

Production Planning and Detailed Scheduling

In SNP, order sequences on machines are not taken into account. Production Planning and Detailed Scheduling (PP/DS) can be used to plan the sequences and to schedule orders (accurate within seconds).

Tasks and goals of PP/DS are the

- planning of the material supply,
- best possible use of scarce resources,
- determination of a sequence that minimizes the setup costs, and
- consideration of unexpected events that might otherwise disrupt the internally determined schedules and the schedules of SC partners.

In particular, the tight connection with ATP allows to specify realistic delivery dates in response to customer inquiries.

PPM use master data that are extracted from both BOM and routings. Activities always relate to a specific resource and can cover several tasks that, for scheduling reasons, cannot be interrupted; examples are setup, processing, and strip down. Procedure selection possibilities ("alternative modes") can be defined for the use of resources. Whereas no historical data are used in PPM, for example, the development of the production master data is also documented over the passage of time in BOM administration systems or PDM systems.

PPM were originally used in the process industries (e.g., chemicals, pharmaceuticals). In order to work with them efficiently in discrete industries (e.g., automotive, PC manufacturers), APO supports integrated Product and Process Engineering (iPPE). The iPPE structure enables rapid line balancing based on capacity constraints. One of the advantages of iPPE is that the APO explosion of the product variant structure (PVS, cf. section 2.1) takes capacity and other constraints (not just material constraints) into account.

SNP passes to PP/DS all orders for which at least one activity lies within the planning horizon of PP/DS. Figure 3.11 illustrates how the data elements that are relevant for planning in SNP and PP/DS differ; a similar observation has also been made in the case of the data used in PP, the production planning module of R/3.

The PPM used in PP/DS is usually defined in more detail than in SNP. This requires all arrivals (from in-house production, external procurement, and restocking) to be rescheduled in PP/DS. For execution, the system automatically plans the order including all associated secondary demands of those parts for which automatic planning was specified in the product master data. As previously explained, the orders realized in PP/DS are displayed in the SNP as aggregated demands for the specified time-buckets.

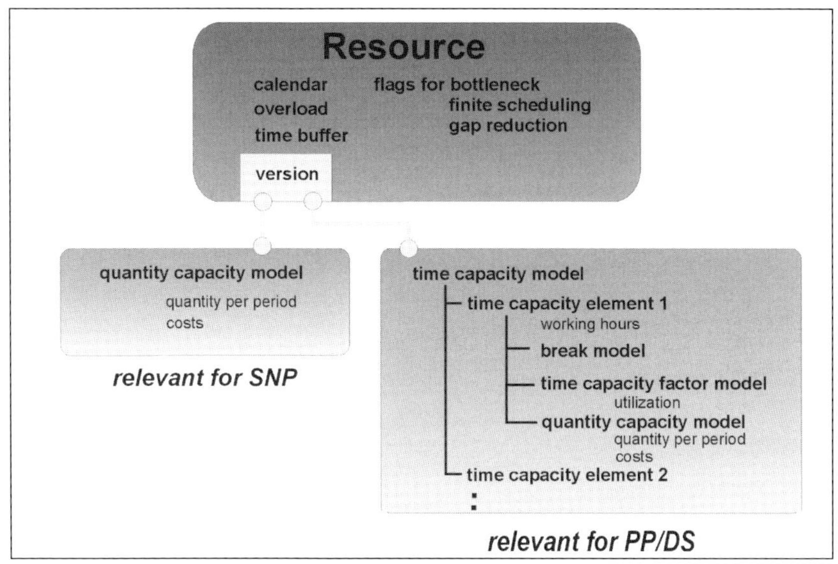

Figure 3.11: Data requirements for applying SNP and PP/DS

PP/DS also includes graphical planning boards similar to control station systems. A planning board automatically displays the current plan version, the status of activities and orders, alternative resources, and planning problems in the alert monitor, plus the activities or orders affected. In addition,

- information about resources, products, orders, and activities
- networks of orders and activities
- information about partial stocks and resource loading, and
- several lists

can be displayed. Furthermore, the planning boards may be used, for example, to activate the following heuristics:

- Clear backlog: Replan all activities that applied to this resource in the past.
- Plan sequence: Selected activities are scheduled into a new sequence in accordance with the settings of the heuristic.

The detailed scheduling strategies for the heuristics are set in the planning boards.

The planning activities based on the board can cover one or more plants, contain a simultaneous material availability and capacity test, perform simulations, and realize the procedure known as pegging; it relates all gross requirements for a part to all planned order releases or other sources of demand that created the requirements. For lower level parts the gross requirements are often pegged to planned orders for higher level items but might also be pegged to customer orders if a part is sold, e.g., as a service part (cf. Vollmann/ Berry/Whybark 1992, pp. 35). The pegging relationships are organized analogously to the BOM structure. Pegging procedures can be used in PP/DS to establish relationships between the receipt (input) and issue elements (output) of a product within a location. If necessary, a demand must be covered with several receipt elements and one production order may cover several customer orders. Thus, pegging can be used to connect a demand with several receipt elements or an order linked with several issue elements. Arcs are used to represent the pegging relationships (Figure 3.12).

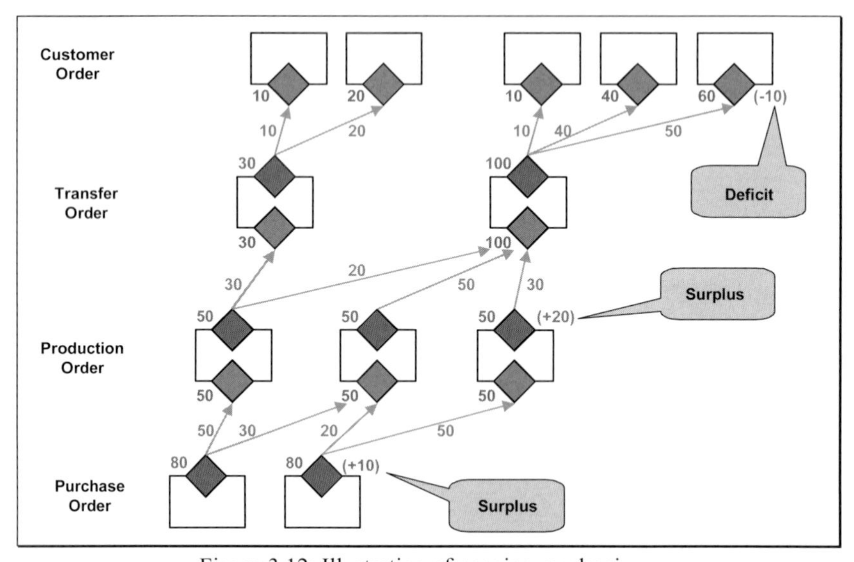

Figure 3.12: Illustration of pegging mechanisms

PP/DS delivers its results to the ERP system or to a Manufacturing Execution System. If necessary, details are added in these systems that were not considered at the APO level; an example could be consumption-controlled C-parts for which no BOM are defined and, therefore, no BOM explosion is performed.

The method can be used to determine production dates that consider the maximum delay of an individual order, the average delay of orders, total lead times, average setup times, and costs of the setup activities. This function evaluates schedules using an objective function which may be a sum of weighted target quantities. The following types of constraints can be observed:

- Working and non-working hours of the resources (calendar)
- Capacities of the resources (finite planning)
- Single or multiple assignment of resources
- Time-related relationships between activities
- Pegging relationships between orders
- Alternatively usable resources from the production process model.

In order to minimize the setup times or costs, setup matrices defined for the associated resources, showing the sequence dependency of setup times and costs, have to be defined and taken into account.

The method determines revised temporal data only for those orders and work steps that lie completely within the planning horizon selected. Scheduling is executed on resources that belong to the propagation area of the planner, have to be regarded in finding a good solution, and have not yet been fixed. All other orders and work steps are fixed, i.e., not replanned. The fixed orders and activities determine by their relationships with the orders and activities that may be rescheduled to what extent the latter can be shifted.

The following parameters must be specified when the PP/DS procedure is called:

1. The acceptable computing time (a goal conflict can arise between the achievable solution quality and the computing time allowed).
2. The planning window that results from the selection of the resources and the time period within which the activities can be rescheduled.
3. The earliest start date at which the orders and activities can be scheduled.
4. Parameters that determine relevant optimization goals and their weights and the solution methods to be applied. These parameters must be specified on a so-called register card.

PP/DS provides the following two heuristics, termed "optimization methods" by SAP:

1. CP Scheduler

   Constraint Propagation (CP) is designed for complex planning problems where many dependencies and constraints need to be considered; therefore, it would be difficult for the planner to find a feasible solution by just using a planning board. CP solves combinatorial problems by quickly restricting the solution space within which the optimum can occur and thus limiting the computing time required for the search. In sequencing, for example, time-related constraints can be used to exclude certain order sequences in advance. If two work steps, A and B, are to be scheduled for machine M, then

   either $\text{begin}(B) \geq \text{end}(A)$ or $\text{begin}(A) \geq \text{end}(B)$

   must hold. If the earliest end time of work step A is larger than the latest begin time of work step B, the sequence A - B cannot be realized and thus the sequence B - A must be chosen. The evaluation of this and other implicit information reduces the combinatorial growth of the number of possible solutions. Nevertheless, because of the computing time involved, determination of the exact optimum is impossible in many classes of combinatorial problems.

2. GA Scheduler

   Genetic Algorithms (GA) use procedures, such as selection, mutation and crossover, that are recognized in evolution in the natural world to find improved solutions for planning problems (cf. Mertens 2000, p. 190). The basic method that uses certain modifications of bit strings has been adapted by SAP for the solution of scheduling problems.

   A typical GA application considers the setup costs in order to determine a suitable sequence of the work steps to be performed. The objective function values are used to select the solutions that appear to offer most potential and should therefore be modified during the search for better solutions. The chances of finding good solutions rise with increasing amount of computing time allocated for problem solution.

During the calculations, the development of the values of the individual goals is displayed graphically. If necessary, additional runs can be made using differently weighted goals. The results of the various optimization runs are stored automatically; they can be displayed and compared at any time. The system accepts the results of the last run and replans automatically as soon as the planner returns to the application from which he invoked the so-called optimization.

Both schedulers provide a similar range of functions. SAP recommends testing the relative suitability of the two methods for the specific circumstances in a certain setting.

Several solution methods implemented in SAP APO use components developed by ILOG. The so-called ILOG Optimization Suite contains three program libraries:

- The ILOG Planner solves linear programming problems by applying the simplex method.
- The ILOG Solver serves as tool to solve the constraint satisfaction problem.
- The ILOG Scheduler supports modeling and provides special algorithms in order to find feasible solutions if constraints exist (cf. Le Pape 1994).

Because different events and resulting update demands can arise during the life cycle of an order, coordination demands occur between PP/DS and the lower level ERP systems. Figure 3.13 shows an example.

3.1.4 Transportation Planning and Vehicle Scheduling

Transportation Planning and Vehicle Scheduling (TP/VS) allows simultaneous consideration of transportation constraints for use of the company's own fleet or carriers for inbound or outbound transportation and for replenishment. The components enable companies, such as manufacturers, retailers, and logistics providers, to liaise via the Internet and to synchronize transportation decisions and activities.

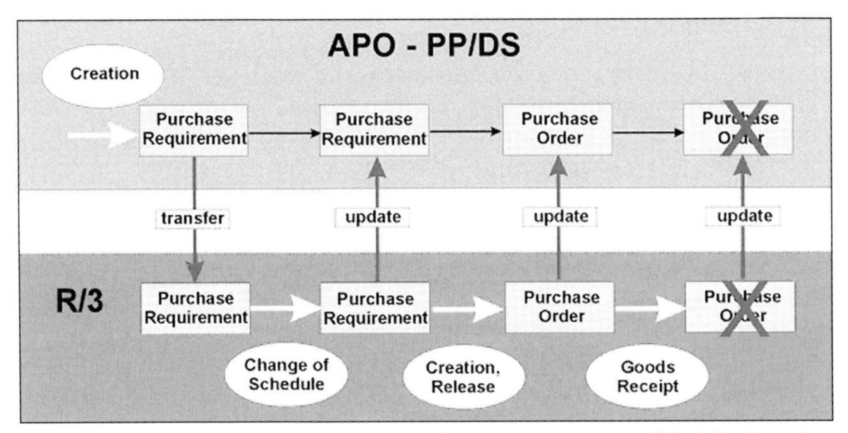

Figure 3.13: Coordination between APO and ERP systems, exemplified for the order management process

Whereas transportation planning focuses on medium- to long-term planning, vehicle scheduling concentrates on short-term planning and routing. Together, they offer the following main functions:

- Load Consolidation and Vehicle Scheduling
- Route determination
- Carrier selection.

Figure 3.14 shows the integration of R/3 SD (cf. section 2.2), LES, and TP/VS (as a component of APO).

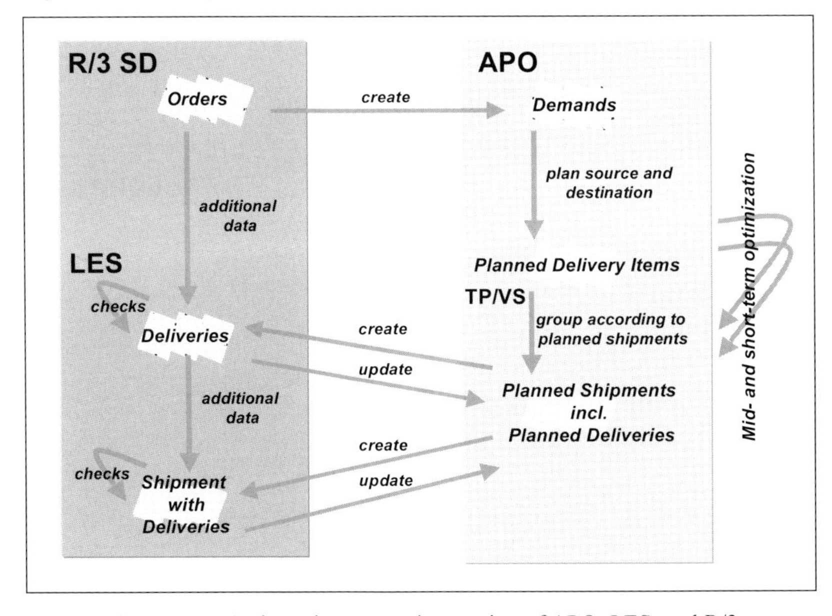

Figure 3.14: Outbound process - integration of APO, LES, and R/3

Load Consolidation and Vehicle Scheduling

This functionality can be used to obtain high filling rates of transportation means. The so-called optimizing engine includes load consolidation for transportation requirements to the destination locations and schedules deliveries to meet the requested delivery dates. It considers constraints, such as conditions for combining different products, special loading requirements, and handling capacities, both inbound and outbound, so that the loading processes at the distribution centers can be improved. Numerous loading processes are defined for different order types. Orders above or below a given size may be automatically released for shipment: these quantities are either too large or too small to legitimate the coordination effort, because the most suitable solution is evident (for example, very small quantities tend to be delivered by a parcels service). All other orders enter the so-called optimization process. The procedures coordinate order grouping, pickup and delivery sequence, and departure and arrival time for vehicles; they also offer alternate solutions.

Vehicle Scheduling, another integral part of the SCM process, focuses on short-term decisions about vehicles, resources, and locations. It addresses pickup and delivery problems, such as how to create the best round trip or whether a trip should include multiple pickups and/or deliveries. The component considers constraints, such as time windows for delivery and vehicles, vehicle capacity, handling capacity for loading/unloading and further checks, including incompatibility checks for certain vehicle/location and vehicle/product combinations.

Vehicle Scheduling requires a supply network planning model that includes a route concept. A route is a sequence of adjacent lanes through a network with a start point and an end point (cf. section 3.1). When a transport request is determined for a route in SNP, this implicitly constitutes a route in TP/VS.

Short-term transportation plans are created by Vehicle Scheduling. The so-called optimizer tries to minimize any existing delivery backlogs as well as the cost of the entire transportation plan within the given constraints, while also allowing different service levels for specific products, brands, and locations. Shipments can be defined either automatically or manually. Any changes made to the transportation plan at short notice result in scheduling difficulties and increased costs for the carrier. These must be weighed against the savings that result from changing the plan. The algorithm is based on the ILOG Dispatcher application for solving a given problem while taking realistic constraints into account, for example:

- Multiple vehicles with different capacity limitations (e.g., weight, cube, pallets, containers, and volume)
- Carrier profile such as partial load [less than truckload (LTL)], complete load [full truckload (FTL)] or equipment availability

- Each stop may have goods to be picked up as well as delivered, though this feature can be deactivated
- Maximum driving time, driver breaks, and other safety regulations
- Pickups and/or deliveries only at a certain time of the day/week (time windows)
- Maximum number of stops per shipment
- Inbound and outbound handling resources at locations.

Operation of the planning procedure is vehicle focused, and the solutions proposed are based on cost comparisons. Customer priorities can also be considered in determining solutions. Carriers often charge different rates for complete or partial loads. Moreover, some carriers offer a discount rate on predetermined tours on which the vehicle movements are prescribed in advance by the carriers and cannot be changed.

Route Determination

A problem frequently encountered in attempts to reduce transportation time and/or costs and improve customer service is that of finding the best route through a network of roads, rail connections, shipping lanes or air routes. The Route Determination component is used to solve this problem. Routes can be associated with one or more transportation orders. Transportation modes, carriers, costs, and schedules can be assigned to these routes. Transportation Planning generates routes automatically or retrieves them from a route library. A route may be a combination of several transports (e.g., a truck from A to B and a cargo flight from B to C) and it is able to serve multiple destinations.

A preliminary decision is concerned with the assignment of destinations to sources. This problem commonly occurs if more than one plant/warehouse can serve a specific location with the same product (cf. Figure 3.15). Scheduling becomes even more complicated when the demand at a certain destination is larger than the production or shipping capacity of the source.

Vehicle movements determined by the so-called optimizer are routes that connect different locations within the supply network. SAP uses the concept of transportation zones to reduce the number of lanes to be maintained in the system and to dynamically generate the lanes necessary for planning. Assume that one distribution center has a transportation lane modeled to a zone with a certain carrier. All customers in this zone may be served by this carrier. The zones may be determined automatically by using the addresses stored in R/3.

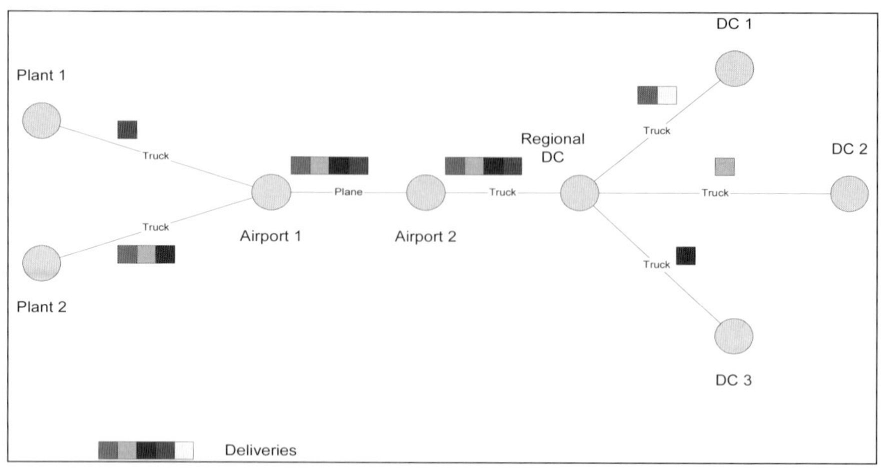

Figure 3.15: Material flow from plants to distribution centers (DC)

Carrier Selection

The carrier profile defined in the master data for Transportation Planning and Vehicle Scheduling specifies vehicle capacities (weight, volume, number of pallets, etc.) and their availability. The planner can choose from several carrier selection processes and decide on the sequence in which they are to be applied. It is also possible to restrict the selection to just some of them.

EDI and the Internet are used to tender a shipment to a selected carrier. The carrier can accept or reject the tender. If the invitation is rejected, a message appears in the Alert Monitor and the transportation planner will either personally or automatically select another one. Alerts may also be generated during manual scheduling (for example, notifying about resource overload) or during data transfer to the execution system (for example, shipment/delivery cannot be realized as planned due to environmental, health or safety restrictions). These messages are displayed in the SAP APO Alert Monitor and may be used as a starting point for replanning or rescheduling.

The Carrier Selection component allows the use of so-called simulation versions and models for what-if analysis. This involves a complex set of calculations performed on reserved data that have been copied from the active TP/VS systems into a special storage area. Once the decision has been taken to use certain simulation results, the parameters of the production systems are modified. After completion of scheduling the results can be transferred to the execution system.

3.1.5 Available-to-Promise

Available-to-Promise (ATP) investigates whether a promised delivery can in fact be made, and if so when. The following aspects, in particular, must be considered (cf. Mertens/Zeier 1999):

1. Determine whether an existing order confirmation can be satisfied with regard to both quantity and date, or check what dates can be promised in response to a particular request.
2. Does the check apply to physically existing and "free" stocks (static perspective) or do planned fluctuations also have to be considered (dynamic view)?
3. Are reservations sacrosanct or should they be changed to favor a privileged customer while disadvantaging others?
4. Is the check restricted to one warehouse location and/or factory or do cross deliveries from other locations also have to be considered?
5. Does the query apply only to end products, or also to the availability of intermediate products over several manufacturing levels through to raw materials obtained from external suppliers?
6. Can or should items not currently available be supplied by the date promised or to be promised?

The various combinations of these issues lead to many different constellations in single companies and SC. As usual, one of the main challenges to packaged software vendors is the need to represent the actual functions, processes, scheduling policies, and priorities in such a way that the characteristics of the company or the SC are supported by the customized software.

The following examples give an impression of the complexity to be handled by ATP systems:

1. An article that cannot be delivered on time may be replaced by a superior one at the price originally agreed ("upgrading").
2. When a reservation is canceled in favor of another customer, the additional profit from the customer order that has been fulfilled must be weighed against any disadvantages (e.g., contract penalties) likely to result from the deferred delivery date for the other order.
3. Inventory may be obtained from another warehouse; however, this results in additional transport costs and may break into decentrally planned safety stocks at the delivering location.
4. Partial deliveries from different locations are bundled to form a complete order, with a consequent increase in costs.
5. Although a part-delivery can avoid too much disappointment for the customer, higher overall costs result.

Figure 3.16 shows an ATP test using an example from the computer industry, in which monitors, desktop cases, keyboards, and accessories are all available in adequate quantities; however, only five motherboards with CPUs installed are in stock. A check of the product inventory shows that no CPUs are available. Consequently, to satisfy the end product demand, at least five CPUs must be supplied and the associated time requirements taken into account when the earliest delivery date is determined.

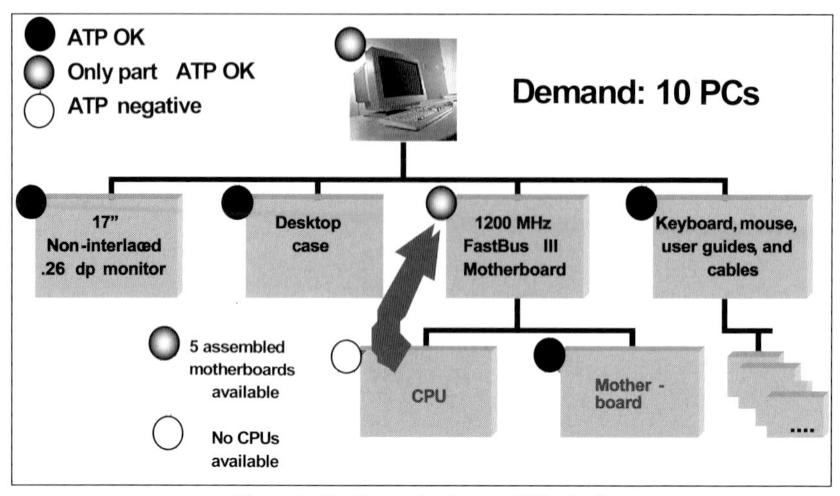

Figure 3.16: Example for an ATP check

Functionality

The special functionality of an ATP system includes:

1. Online queries concerning availability by the inventory planners using various databases.
2. Cooperation using various packaged and customized software modules of partner companies in the SC.
3. "What if" or "How to achieve" analyses to test alternate solutions that could be used for fast supply of the parts.
4. Assignment of dynamic priorities for orders, customers, warehouse, and transport alternatives.

Because an exact optimum can be determined with an acceptable amount of effort only in exceptionally simple situations, simulations, heuristics, and knowledge-based rules that may also include an explanatory component can be combined.

SAP provides several procedures for ATP checks as part of APO. The ATP Decision Cube contains rules specifying which alternatives (e.g., delivery of a substitute product, choice of the second-best location or production of an out-of-stock article) have to be tested in what sequence. Figure 3.17 shows, as an example, an extract from an ATP rule: In this case, the system first tests whether an alternate product is available at the same location. If not, it tests whether the same product can be obtained from another location. If the answer to this query is also negative, both deviations may occur simultaneously. Finally, the only remaining option is to produce the required product because no stocks can be assigned to the order under scrutiny. In this case, the test continues by determining which BOM items needed for this product are in stock. The test logic may have to be repeated at several manufacturing levels.

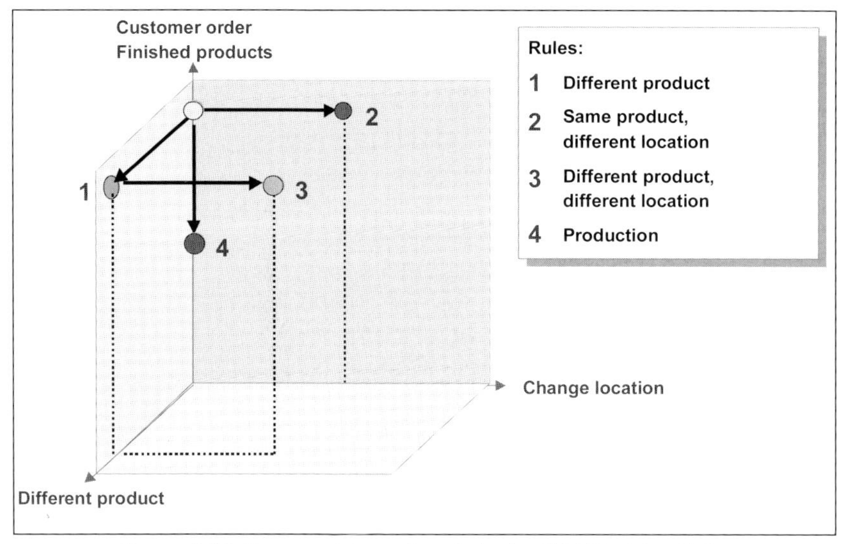

Figure 3.17: Representation of a rule in the ATP Decision Cube

If a confirmed delivery date cannot be met, APO allows rescheduling of the order. The high priority of this order will be taken into account by all subsequent planning processes to ensure that the customer demand is met as soon as possible.

Global ATP and Local ATP

SAP makes two variants of the ATP available:

1. *Local ATP* permits the ATP check to run on a single R/3 System. This version is provided as a module in *one* R/3 environment.
2. *Global ATP* provides the general capability to run the availability check simultaneously on several R/3 Systems, on products from other packaged software vendors, and on legacy systems.

Because worldwide delivery networks require, for example, that inventory, production and transport capacity of several independent companies must be checked, they represent big challenges for ATP checking. The production plans of partners can be either accessed in real-time or as derived and, thus, less up-to-date data. Furthermore, these data can be accessed at the partner's system or an extract of the relevant data can have been transferred. Special technical defiances arise from the high-availability requirements of ATP: The APO and its interfaces to R/3 must eventually be available in a 7*24 hour operation and the relevant data may have to be mirrored, e.g., for allowing system maintenance.

3.2 Business-to-Business Procurement, EnterpriseBuyer

SAP Business-to-Business Procurement provides electronic support for the procurement of goods and services and for integration of the purchasing process into the overall flow of goods, information, and finances. Recently, SAP has presented this concept under the name of mySAP EnterpriseBuyer [http://www.sap.com/solutions/scm/scm_over.htm]. However, we use the established terminology below.

Business-to-Business Procurement was developed especially for the purchase of C-parts by members of staff to whom a separate purchasing budget has been allocated ("indirect procurement", "MRO materials management"; cf. section 2.3.4). The component provides employees with the capability to procure goods and services independently during a session at their own workplace, based on the Internet technology. The operative tasks are passed to the individual member of staff, and the purchasing department can concentrate its efforts on more strategic decisions (e.g., supporting the choice of suppliers, negotiating terms of delivery and outline agreements). This procurement task can be embedded in a workflow that allows for involvement of managers for certain types of purchases.

3.2.1 Web-supported Autonomous Procurement

Business-to-Business Procurement uses Internet technologies to permit interactions between purchasers and suppliers. The following aids are provided:

1. Create and change reservations, order requests, and orders with or without use of a catalog.
2. Approve or reject order requests; the web-capable SAP Business Workflow permits control and monitoring of the procurement process.
3. Because the user can confirm the arrival of the material requested without involving a central unit, goods receipt is streamlined.
4. Because the sales department of the supplier can derive invoices via extranet from the order data and pass them on to the purchasing department or to other responsible units of the customer company for approval, invoicing and payments are also simplified.
5. Service providers, such as consultants or freelancers, are able to enter their data in the Business-to-Business Procurement system directly so that the work done can be paid for without delay.
6. Electronic placement of orders can be available only to well-known suppliers (restricted tendering) or publicly accessible to any interested party (public tendering). The business partner who offers the lowest price for the product or service or who provides other advantages wins the order.

Figure 3.18 illustrates the process of procuring C-parts. The component supports these actions using a web-based front end that permits call up of the relevant functions from the Business-to-Business Procurement start page (Figure 3.19). Figure 3.20 describes the tasks of the user groups involved in this process.

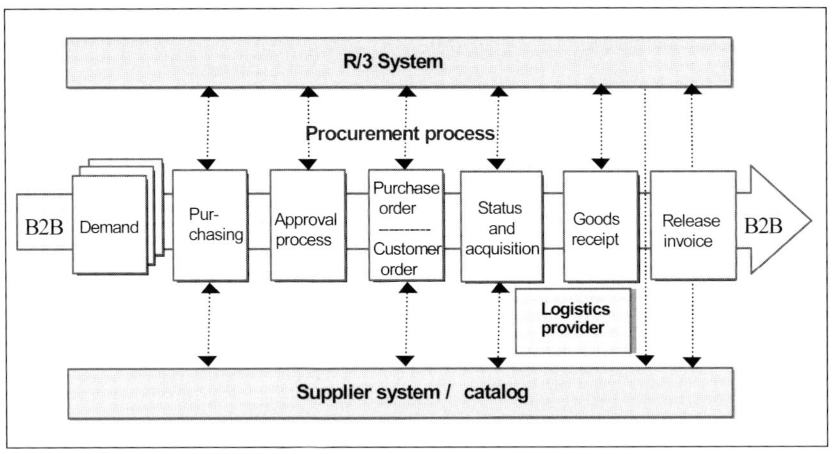

Figure 3.18: Procurement of C-parts via the Internet

3.2.2 Connecting Catalogs

An essential function of Business-to-Business Procurement is the availability of flexible connections to product catalogs. The SAP solution supports three types of catalog connections:

1. The administration of the company's product catalog
2. The integration of catalogs from various suppliers using a catalog broker
3. Direct access to the suppliers' web catalogs.

Catalog brokers are companies that combine product directories from various suppliers to form a "virtual catalog"; they free buyers from searching web-sites of various suppliers (cf. section 2.3.4).

The direct catalog access permits a B2B interaction between a purchaser and a supplier. When using the „Online Store" (cf. section 2.3.4), SAP Business-to-Business Procurement creates a sales order for the supplier and a procurement order for the customer. BAPIs provided by SAP ensure the ability to combine SAP Business-to-Business Procurement with several packaged and customized software systems.

Figure 3.19: Start page for Business-to-Business Procurement

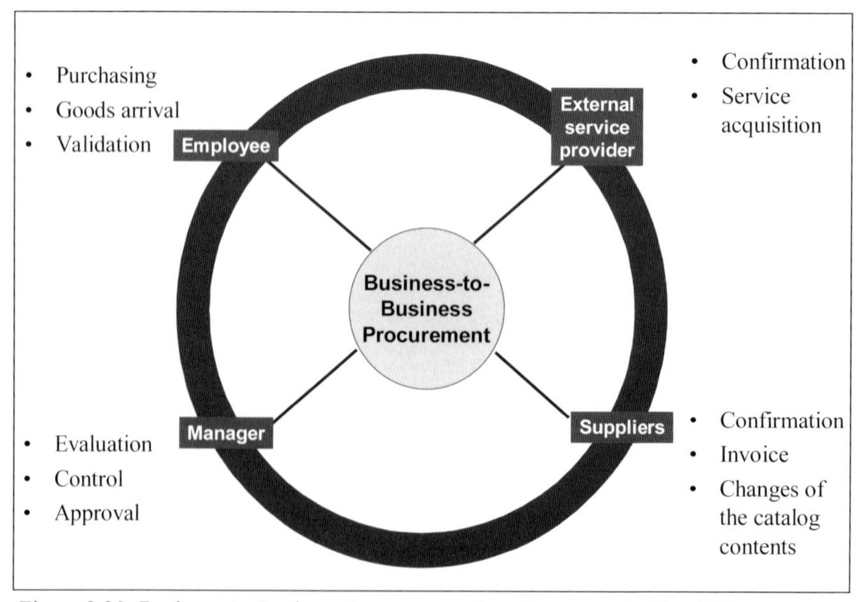

Figure 3.20: Business-to-Business Procurement functions provided for different groups

3.3 Logistics Execution System

Whereas the APO system provides functions for demand and production planning and for an availability check over all nodes of a logistical network, the Logistics Execution System (LES; cf. Buxmann/König 2000, pp. 109) supports economical warehousing and distribution based on the Warehouse Management System (WMS) and the Transport Management System (TMS). LES allows access to all base technologies, such as master data administration used in the R/3 System, and is fully integrated with the APO components TP/VS and ATP.

3.3.1 Warehouse Management System

Although the WMS is a component of the LES, it can also be used as a standalone component. Figure 3.21 illustrates the WMS functionality. In addition to the SAP interfaces, the WMS provides interfaces to other ERP systems, picking systems (e.g., Pick-to-Light), and warehouse technology, administration and control systems. WMS supports

- the management of warehouse structures and equipment (e.g., automated warehouses, high rack storage areas, bulk storage or fixed bin warehouses),
- the overview of warehouse movements,
- the management of the activities that accompany the goods' receipt and issue (e.g., unloading, packing goods, dispatching delivery vehicles to the appropriate loading ramps, and print shipping documents),
- the maintainance of the current inventory data for storage bins using continuous stocktaking,
- the putaway and picking of dangerous goods and all other materials that need special handling, and
- special services, e.g., customized packing and labeling procedures.

The use of wireless terminals and mobile barcode scanners that are directly connected to the WMS simplify the warehousing processes. SAP provides a tool that meets this need, the integrated Radio Frequency (RF) component. It uses mobile RF terminals with scanning devices to enable immediate data transfers. RF terminals can be mounted on forklifts or in hand-held units and receive data directly from the R/3 System.

One can record information, such as storage bin coordinates or EAN article numbers, by scanning a concatenated barcode (based, e.g., on UCC/EAN128) and then use that information to verify the current storage location information. SAP's standard graphical user interface displays information on graphic terminals. For character-based terminals, SAP offers an interface, the SAPConsole, which converts the SAP standard online transactions into a character-based ASCII user interface.

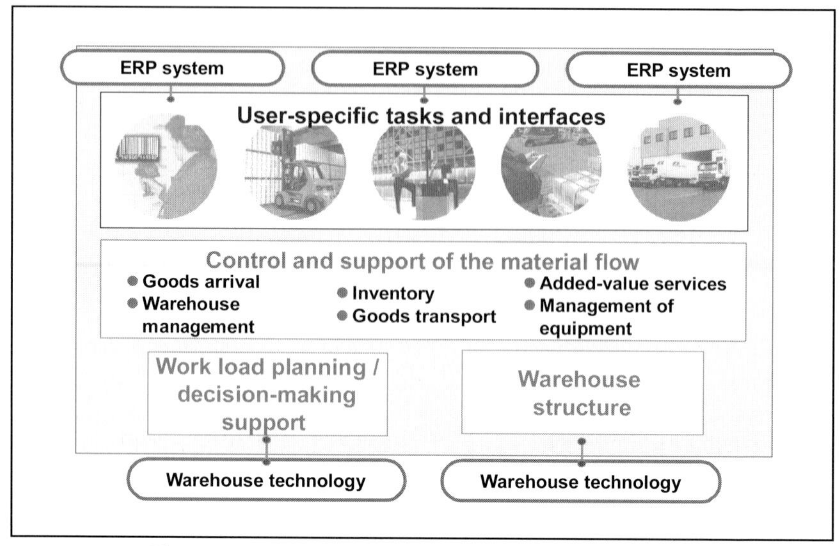

Figure 3.21: Warehouse Management System of the LES

The RF solution contributes to improved

- data collection with EAN128/SSCC barcodes
- stocking, picking, and transfers
- packing
- ship/load control
- displays of stock data
- monitoring functions.

The MOB-to-R/3 product from WITRON [http://www.witron.de/] which is used, among others, by Roche [http://www.roche.com/], provides another example for a radio-controlled communication and mobile data acquisition in warehouses.

Both production planning and detailed scheduling are based on data from the R/3 Logistics Information System (LIS). It permits a workload preview, among other things; if bottlenecks are expected, additional employees can be assigned temporarily to the workplace affected. The Warehouse Alert Monitor sends warnings about processes that have not been completed on time. This tool is similar to the Alert Monitor in the APO system and can also be customized.

The WMS also supports the management of handling units (HU). An HU consists of packaging material and the materials contained therein. Once an HU has been defined, it will be used to execute goods movements for the total content in a single step. HU also simplify communication with SC partners over the storage, distribution, and transportation processes.

3.3.2 Transport Management System

The transportation execution uses the plans determined in TP/VS to provide functions for scheduling shipping, for route planning, for calculating the freight costs, and for transport handling and monitoring. There are also interfaces to non-R/3 Systems (e.g., to so-called route optimization programs and distance databases). EDI systems and Internet services are available as communications media with customers and suppliers (cf. Figure 3.22, using the acronyms CLM for Car Location Message and GPS for Global Positioning System). SAP has just developed an SC tracking system (cf. section 2.5.3).

Scheduling of orders and route planning must be performed before a transport order leaves the site. The TMS serves as the basis for selection of service companies, assignment of a carrier, determination of appropriate transport means, identification of the transport route and the associated lanes, and scheduling the shipping. The so-called TMS Optimizer can be used to plan the transport in simple cases while observing the constraints specified during customizing. Examples are a weekly delivery of specific products and the transport of dangerous products exclusively by rail.

At the beginning of 2001 it was impossible to perform transport planning using the optimization procedures in the APO method library. However, TMS contains standardized interfaces that can be used to include travel improvement programs from other vendors. In this area, certified SAP partners include CAPS Logistics, Dr. Städtler, InterTrans Logistics Solutions, McHugh Software International, PTV Planungsbüro Transport und Verkehr, and Soloplan. The TMS provides many functions to calculate the freight costs. Examples are:

- Acquisition and administration of master data (e.g., conditions, means of transport, tariff zones)
- Assignment of service providers (e.g., carriers) to a transport and calculation of the associated transport costs
- Determining the transport costs per travel section taking account of basic freight, negotiated profit margins, surcharges, etc.
- Settling up with the service providers and notifying the billing system of the transport costs
- Billing of the transport costs to external or internal customers.

The R/3 functions for pricing have been extended to allow their use for consideration of the freight costs. The central instrument is a calculation scheme that includes a list of all condition types used to estimate those costs. A shipment cost document is available for calculating a transport service. This document contains information about the settlement data, the basic data used to compute the costs, and the result of this computation. Freight costs can be calculated based on weight, volume, distance, and customer-specific parameters.

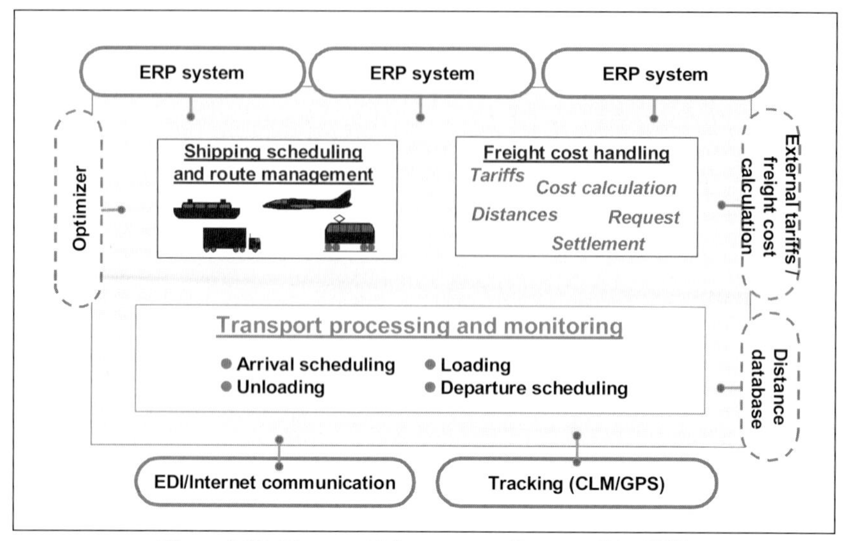

Figure 3.22: Transport Management System of the LES

As in shipping and route planning, standardized interfaces to third-party systems permit the calculation of complete freight costs or individual cost components (e.g., base freight, negotiated discounts and surcharges). Examples of such systems are Venture Freight Management from InterTrans Logistics Solutions and TMS from McHugh Software International. The interface permits transfer of the amounts calculated from non-R/3 Systems to R/3 where they can then be processed further.

By defining start and end dates, SAP TMS supports the processing and monitoring of the physical transport system. For example, a transport activity receives the status "completed" only when the date and the time of the delivery have been recorded.

There are standard reports, e.g., for transport scheduling or transport completion; the route of an individual transport can also be displayed graphically. Furthermore, the TMS reporting system is also connected to the BW which may be used to create additional reports. For example, it can be determined how many deliveries a specific carrier handled the previous month.

3.4 System Architectures

3.4.1 Architecture Variants for Hierarchical Planning

The cooperation between SCM and ERP systems is a form of hierarchical planning (cf. Hax/Meal 1975; Dempster et al. 1981). In this concept, a comprehensive planning model which because of its complexity cannot be simultaneously optimized is decomposed into (usually two or three) hierarchically arranged models of restricted scope. Optimization methods or heuristics can be used at the individual planning levels in order to find optimal or good solutions for the subproblems. The preliminary solutions are exchanged between the different planning levels. This results in an iterative procedure for which, in particular, the following termination criteria may be defined:

- The result of the subproblem remains unchanged with regard to the objective function value determined in the previous iteration.
- The objective function value of the solution of the subproblem has improved by less than p percent compared with the result from the previous iteration (e.g., p=0.5%). A special case is p=0, i.e., the value has remained unchanged.
- The maximum number of iterations has been reached.
- The time when the solution must be available does not permit further computations.

The methods used at the individual levels of the hierarchical planning process must be customized for the specific market and production conditions. For example, linear programming can be used to determine optimal production programs for part-families at a highly aggregated level employing estimates for setup times from historical data. After the optimum solution to this highly simplified model has been passed down to the lower level, it can be determined heuristically what products are to be manufactured in what quantities and in what sequences. On this basis, the total setup time associated with the production program determined at the upper planning level can be calculated. A check is then made on whether this total agrees sufficiently well with the capacity demand that was deducted as estimated setup time at the higher level. The solution will be accepted if the difference lies within the tolerances specified by the planner. Otherwise the capacity allocated for setup is modified and the linear program is rerun.

Figures 3.23 a-d show various conceivable IS architectures supporting such a hierarchical planning process. It must be clear whether the systems that exist at the upper and lower levels of this architecture communicate with each other or whether they operate horizontally in isolation. In case 3.23 d it is not necessary for the number of systems implemented to be the same on each of the two levels.

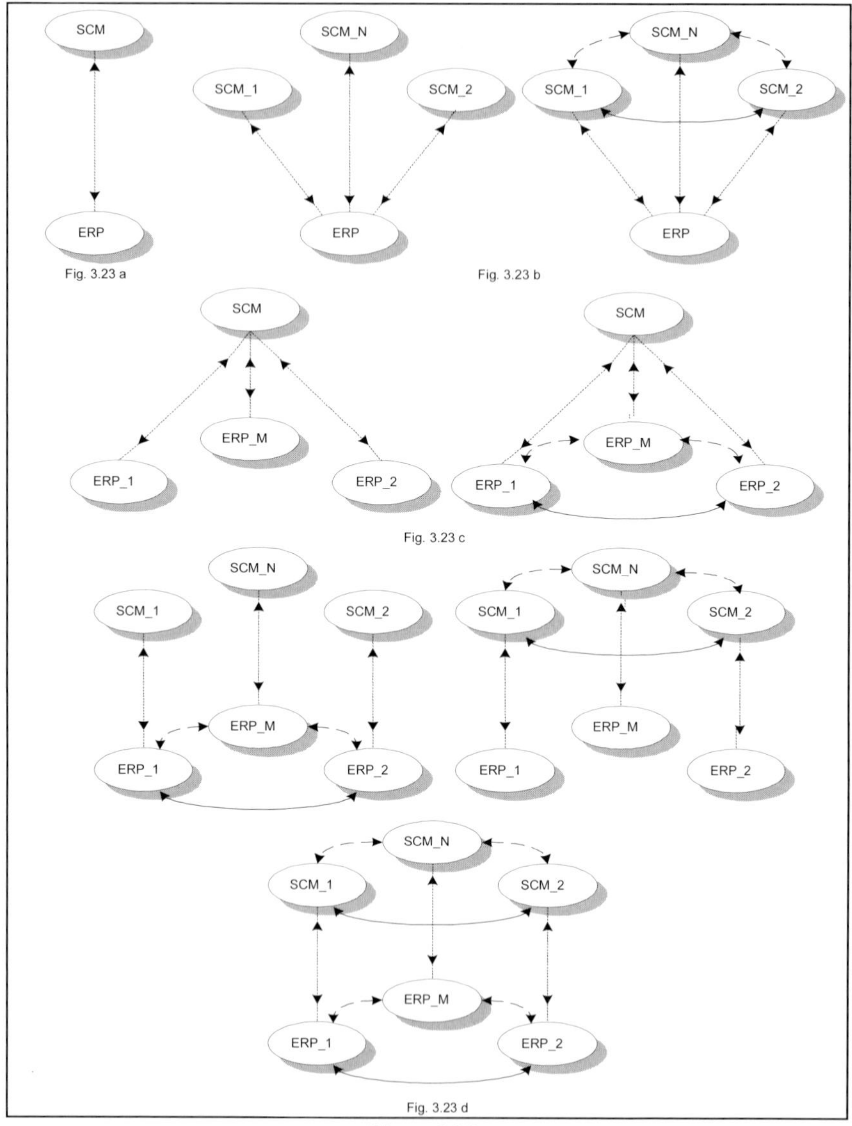

Figure 3.23

3.23 a: Coordination between one ERP system and one SCM system

3.23 b: Coordination of one central ERP system with several SCM systems

3.23 c: Coordination of several ERP systems with one central SCM systems

3.23 d: Coordination between several ERP systems and several SCM systems

In SAP environments, the following relationships may exist between R/3 and APO systems:

- 1:1 relationship: One APO system is assigned to each R/3 System as a component designed to realize advanced concepts. This may apply, for example, to Local ATP. The architecture for the hierarchical planning concept shown in Figure 3.23 a is exemplified in Figure 3.24 for the SAP systems R/3 and APO.
- 1:N relationship: The master data for a group's production planning are administered in a single R/3 System; one plant-specific APO system supports each decentralized plant in its local planning and scheduling tasks. A similar architecture is realized, for example, by the German fischerwerke group (cf. section 5.1.2).
- M:1 relationship: One R/3 System that uses the PP module exists in each individual plant within the SC. APO is used for cross-plant coordination within the SC (cf. Figure 3.25 and the Colgate-Palmolive case study in section 5.1.1). This architecture comes closest to the theoretical concepts of hierarchical planning and SCM. A generalization of this concept is a central node which serves as the SCM hub (cf. Figure 3.26).
- M:N relationship: Several decentralized R/3 Systems and several coordinating APO systems exist within a group. As described above, M>N, M=N, and M<N may hold. From the viewpoint of hierarchical planning, M>N seems to be the most plausible.

What all the variants described have in common is that the APO components access databases of the R/3 Systems and use selected, possibly aggregated, data for higher level planning purposes. The R/3 Systems realize the lower level planning and scheduling processes for the plants to which they are assigned.

In group companies (cf. the Wacker Siltronic case described in section 5.1.4) in which R/3 is used both centrally and decentrally, APO can serve as a coordinating layer between these systems (cf. Figure 3.27). Figure 3.28 illustrates the information flow between R/3 and APO for a components supplier in the automotive industry. The analysis of the connection possibilities between SAP R/3 and APO must also consider industry-specific systems as implemented in SAP Industry Solutions.

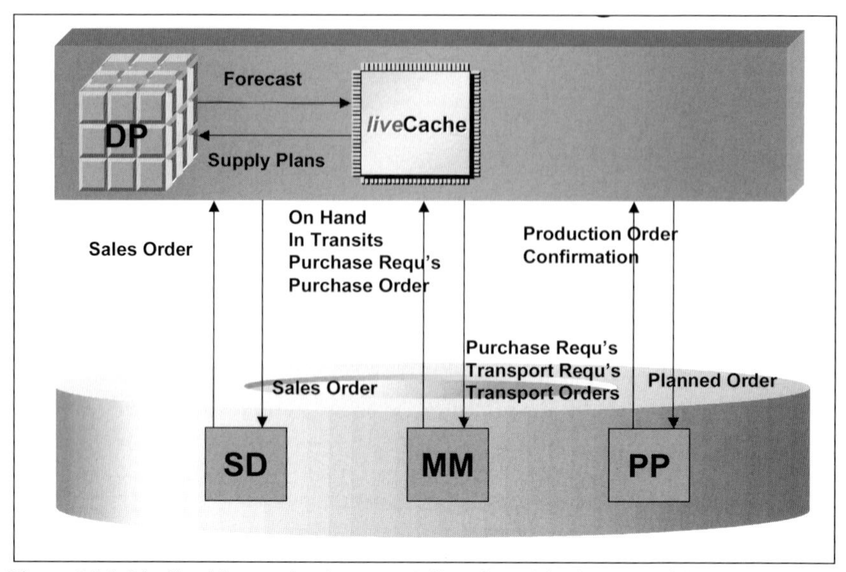

Figure 3.24: Idealized interaction between R/3 and APO in the case of demand planning

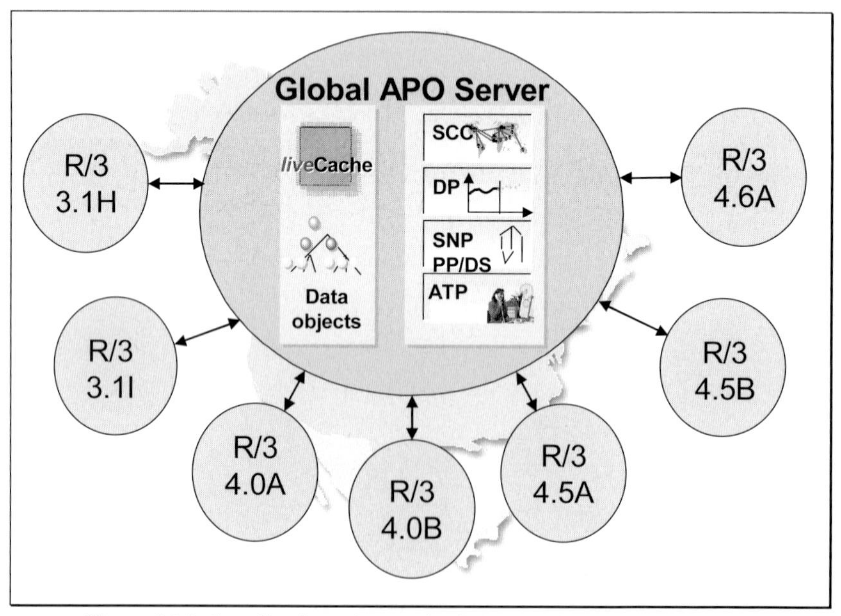

Figure 3.25: Coordination of several R/3 Systems by one APO System

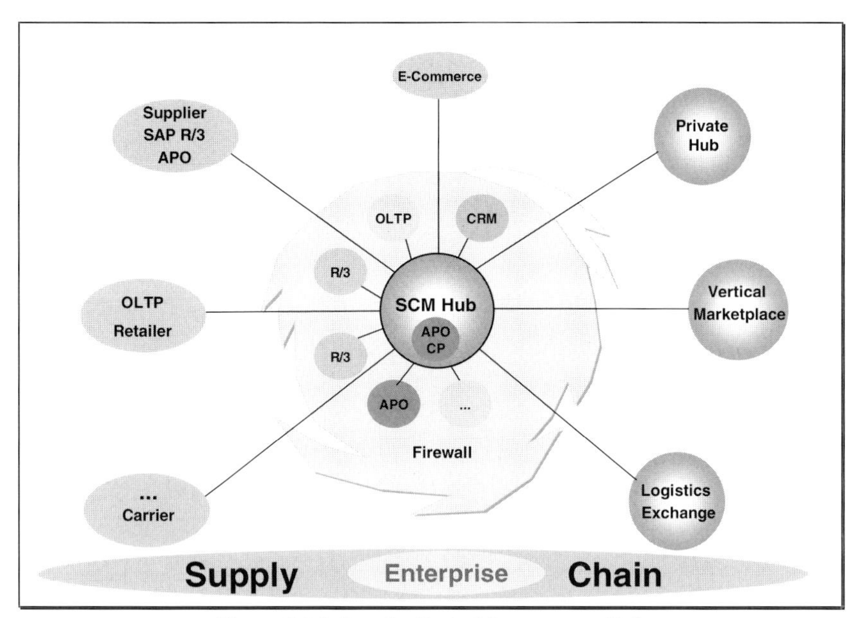

Figure 3.26: Supply Chain Management Hub

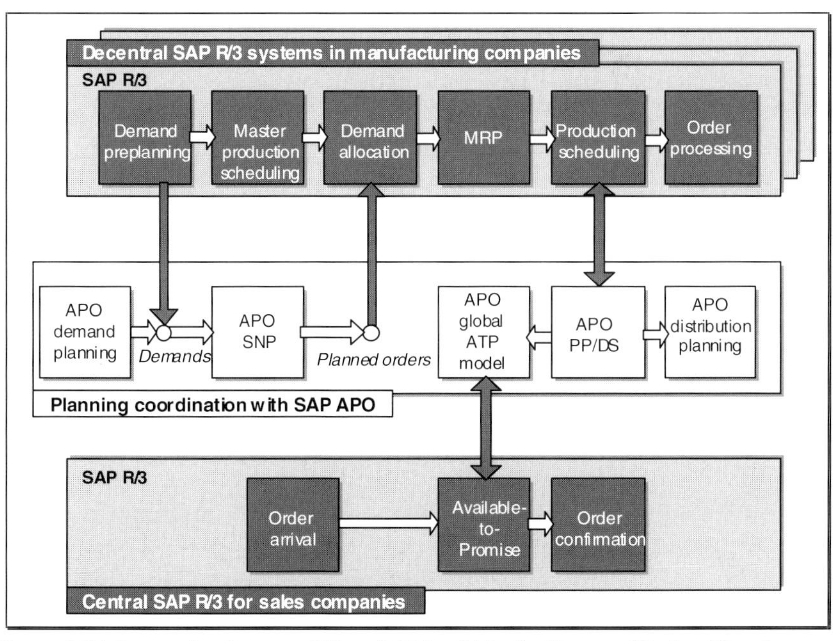

Figure 3.27: Interaction between R/3 and APO within the German Wacker Chemie group (Hurtmanns/Packowski 1999, p. 67)

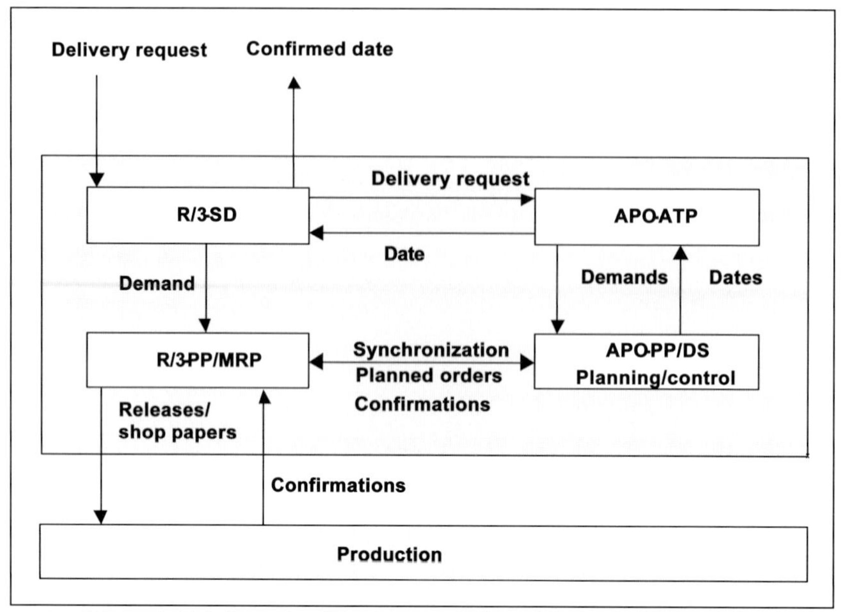

Figure 3.28: Information flow between R/3 modules and APO modules (cf. Bothe 1999, p. 74)

The APO Core Interface (CIF) is responsible for the data exchange between APO Systems and R/3 Systems (cf. section 3.4.2). The connection of APO to non-SAP systems can be realized using the BAPIs described in section 3.2.2 (Figure 3.29).

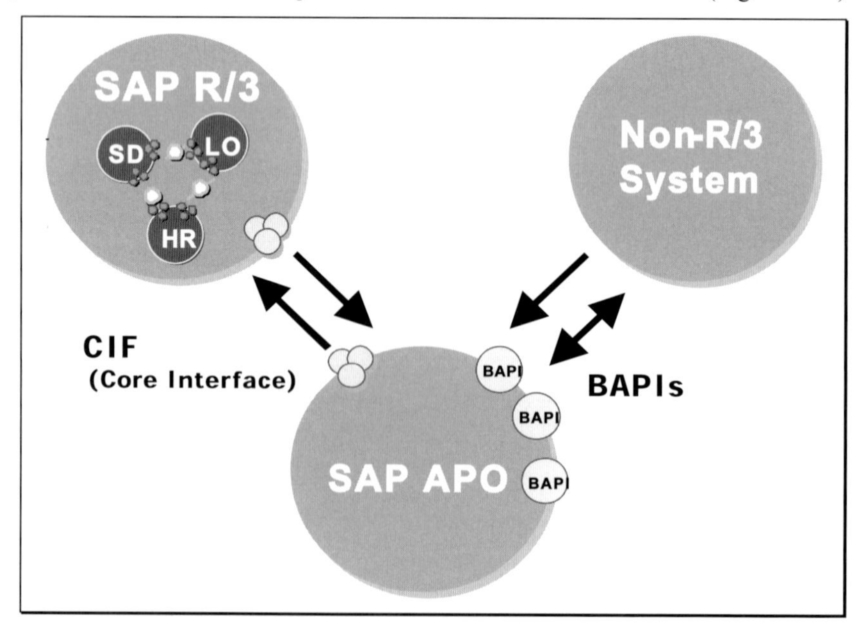

Figure 3.29: Data exchange between APO and R/3 and between APO and non-R/3 Systems

3.4.2 Supply Chain Management in Homogeneous SAP Environments

The APO Core Interface (CIF) allows data exchange between the R/3 and APO systems. For this purpose, SAP provides a supplementary program that deactivates specific functions in R/3 and activates similar, but more powerful, functions in the APO. When both SAP systems are used, the Core Interface permits a comprehensive integration that can be used, for example, to perform the BOM explosion either in R/3 or in APO.

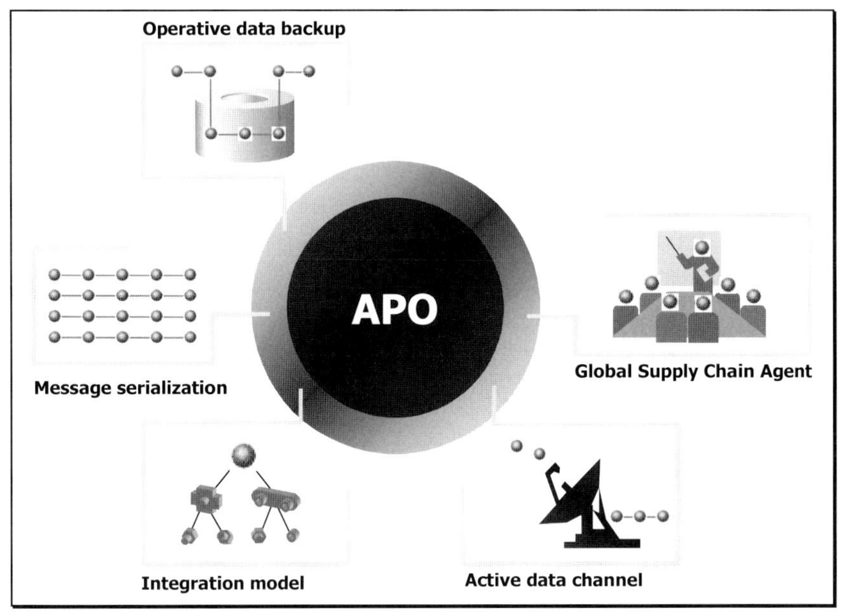

Figure 3.30: Functions of the APO Core Interface

The CIF includes five functional features (Figure 3.30):

1. Integration model

The integration model determines the type and size of the objects (e.g., material stock, BOM, routings, purchase orders) for which data are to be transferred to the APO. Different integration models can be defined for diverse forms of division of work between R/3 and APO. Each model defines what data are to be passed between the systems or what types of data are to be processed in the APO and then passed to R/3 for further processing. SAP provides a model generator as part of the APO CIF (Figure 3.31).

2. Message serialization

This component ensures the referential integrity between databases used by R/3 and APO. Apparently there are logical dependencies between the elementary data in R/3 and the (possibly aggregated) data managed in the APO. To minimize the communication volume, CIF transfers only those data changes that are relevant for the partner system. To ensure integrity, these changes must be performed in a specific sequence (serialization).

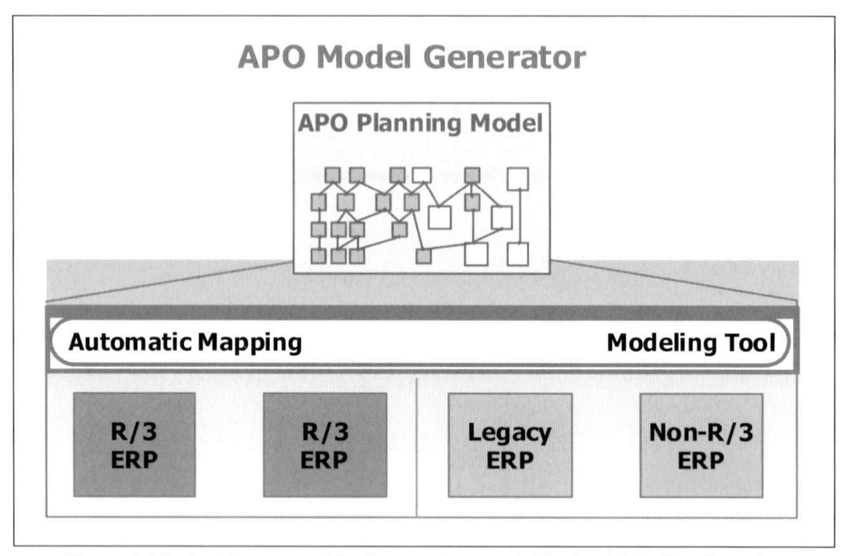

Figure 3.31: Architecture of the integration model in the APO Core Interface

The data are transferred using Remote Function Calls (RFC). Specifically for the APO, this call was extended to become a transactional procedure, ensuring that the data are processed in the destination system in the correct sequence. This is particularly relevant for time-critical transaction data (e.g., shipping quantities), which must be handled close to the time of data creation. The processing sequence is not determined via time-stamping but by centrally assigned, increasing numbers.

3. Operative data backup

It is necessary to check the integrity of data so that in case of any problems the possibility that these are caused by the forwarding system can be excluded with high probability. CIF provides a monitoring function in order to track the time and source of errors. The Operative Data Backup component can also be used for recovery purposes.

4. Active data channel

The activation of a previously generated integration model triggers the initialization of the APO with data ("Initial Data Transfer"). Simultaneously, the event-driven transfer of "delta data" is also activated ("Incremental Data Transfer"; cf. Figure 3.32). Parallel processing is realized because only subsequent changes to the same object will be serialized, whereas modifications of different objects can be processed simultaneously. The transfer of the relevant data follows the push principle, in which changes initiate one or more data transfers immediately or following well-defined rules.

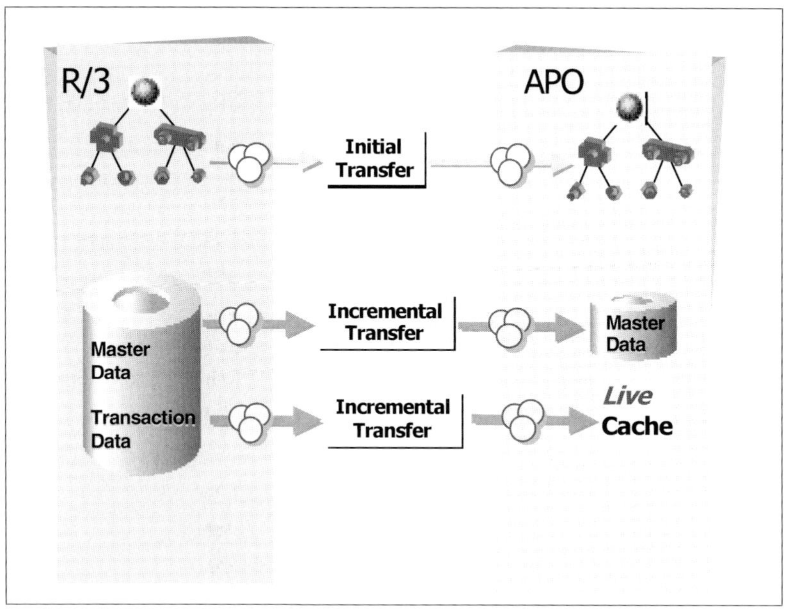

Figure 3.32: Data transfer via the CIF

5. Global Supply Chain Agent

This function controls the distribution of data to and from the connected R/3 Systems. The corresponding target system can be identified by using the locations of the APO order types. Some of the candidates for data forwarding are planned orders and production orders from the SNP and PP/DS modules, purchase orders, customer orders, and stock transfers.

3.4.3 Supply Chain Management in Heterogeneous SAP Environments

SCM on the basis of SAP systems is realized not only in homogeneous SAP environments but also in heterogeneous environments. Basically, four variants are conceivable:

1. Interaction of SAP APO with non-R/3 Systems
   1.1 exclusively non-R/3 Systems
   1.2 both R/3 Systems and non-R/3 Systems
2. Interaction of SAP R/3 with non-APO systems
   2.1 exclusively non-APO systems
   2.2 both APO systems and non-APO systems.

Heterogeneous environments result if no stringent IS strategy is followed, if decisions about IS architectures are highly decentralized, or if mergers and acquisitions have contributed to the diversity. For example, the Swatch group, in which several formerly independent companies were merged, is using more than five different ERP and MRP II systems. Obviously, heterogeneity of IS environments leads to additional problems in interfacing SCM and ERP or MRP II systems.

Depending on the system architecture, either BAPIs or specially developed SCM interfaces (e.g., the Production Optimization Interface; POI) are available for data transfer in heterogeneous SAP environments. Thus, SCM is supported using an existing infrastructure. The following issues may cause problems:

1. Conventional interface problems
2. Insufficient synchronization of the data models of various software systems
3. Batch-oriented data transfer
4. Limited functionality (e.g., no real-time planning) of some components.

Interaction of SAP APO with non-R/3 Systems

The APO accesses databases of the ERP or legacy systems and uses selected, possibly aggregated, data for high-level planning purposes. The functionality of APO is independent of whether R/3 or another system is applied at the lower level. The BAPIs available for data transfer are SAP-specific, object-oriented Application Program Interfaces (APIs) that support the interaction between SAP and non-SAP systems. BAPIs permit APO systems to access databases of the lower-level execution systems and to transfer APO data to them.

Furthermore, BAPIs allow front-end applications an access to the Business Object Repository of the R/3 System. BAPIs can also be used to access certain functions of the R/3 System via the use of Internet Application Components (IAC) and an Internet Transaction Server (ITS).

Interaction of SAP R/3 with non-APO systems

The BAPIs described above can be used for this data transfer. The R/3 System also provides four interfaces to support an SCM system:

1. Production Optimization Interface (POI)
2. Demand Planning (DP) Interface
3. Distribution Requirements Planning (DRP) Interface
4. Transport Planning (TP) Interface.

The first interface is of particular importance, because it can be used to connect special SCM tools with the R/3 System. Some certified tools that are able to communicate with R/3 are, for example, RHYTHM from i2 Technologies or Numetrix/3 and Numetrix/xtr@, now offered by J.D. Edwards. In particular, the POI permits the transport of master data (material, BOM, routings, etc.) and transaction data (e.g., planned orders, production orders, demand lists, and inventory lists) from an R/3 System to an external planning system. After the external planning has been completed, any changes (e.g., in planned orders and production orders) can be forwarded to the R/3 System.

An example for realizing the SCM concept via the Internet is the "B2B Relationship Management" software offered by Extricity Software, consisting of the

- Extricity B2B Alliance Manager
- Extricity B2B Integration Adapters
- Extricity B2B Partner Channels
- Extricity B2B Process Paks [http://www.extricity.com/].

Information is transferred among several companies in star fashion using special servers. On the one hand, connections can be established to the products from the leading ERP vendors and, on the other hand, to various types of middleware. Integration Adapters provide the communication bridges for supporting systems such as SAP, Baan, Oracle, PeopleSoft, i2, and messaging middleware. Figure 3.33 illustrates the architecture of a solution previously described by Extricity as "Any-to-Any Application Integration." The implementation of the SCM solution requires close cooperation between the SC partners, because the existing IS architectures and the customization of packaged software systems also need individual solutions for system integration.

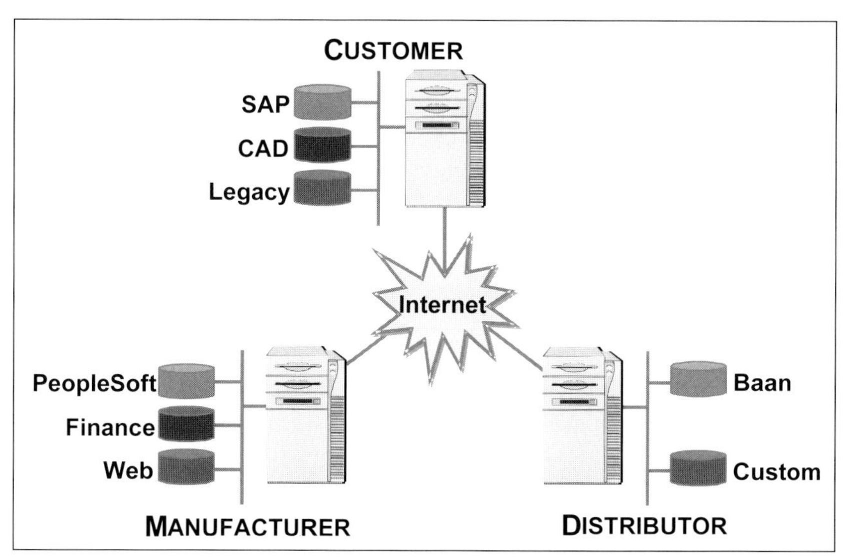

Figure 3.33: Integration of different systems in the SC using Internet technology (Extricity Software)

4 Recent Developments in Order and Supply Chain Management based on SAP Systems

The highly dynamic situations in the development of order management, SCM and the architecture of large information systems, both in practice and in the academic world, will lead in the next few years to many new ideas, methods, standardization concepts, and software products. The following discussions are based on information obtained from SAP, market research, personal conversations, press releases, and Internet sources.

4.1 mySAP.com Solutions - Effects of the Internet on the SAP Environment

The Internet is increasingly important for order management, in both sporadic and permanent business relationships. This has substantial effects on ERP systems, which are used in many companies as IS backbone for the associated transactions (cf. Hantusch et al. 1997). Consequently, all ERP providers, and especially SAP, are undertaking huge efforts to utilize the many developments on and around the Internet and to make their software solutions web-compatible. The SAP AG is positioning itself, in particular via the mySAP.com approach, as an Internet company. This focus first became visible in new user interfaces known as EnjoySAP in release 4.6 of SAP R/3. The potential of the Internet also changed the internal architectures of ERP systems.

mySAP.com encompasses an enterprise-wide portal (mySAP Workplace) and a marketplace. Both serve as a platform for other systems within the mySAP architecture (cf. Figure 4.1).

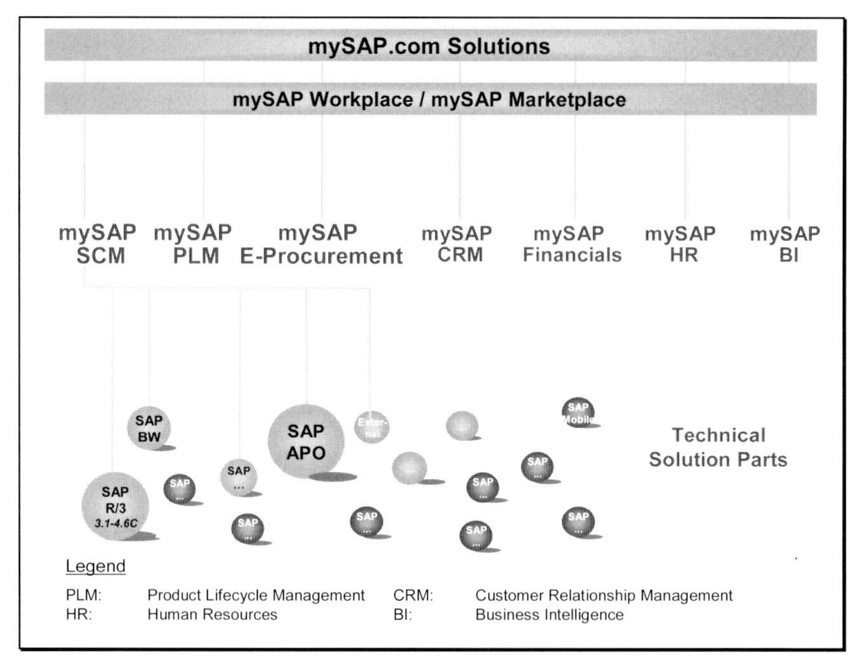

Figure 4.1: Architecture of the mySAP.com solutions

- mySAP Workplace

  is an enterprise portal providing users with a central access point to information, EDP applications, and services. Members of staff can use the mySAP Workplace systems from their workplace in the firm, from home, and also from a hotel, using a role-based and personalized interface.

  Figure 4.2 may give an impression of how staff members, using the portal, get in contact with several other tools and partial systems of the "SAP world".

  SAP hopes that the majority of its more than 20,000 customers will provide information via this portal. Another software vendor, called webMethods [www.webmethods.com], offers specialized XML tools for the integration of the mySAP portal with components of other software vendors. As an example, they may be used to connect mySAP SCM with the MRP system of a supplier.

- mySAP Marketplace

  was developed by SAP's daughter company SAPMarkets in cooperation with Commerce One. It extends the boundaries of SCM to include suppliers, manufacturers and retailers and enables collaborative business processes. Whereas without the mySAP Marketplace the APO of one company can only exchange data with another APO (of the partner company, cf. section 3.4), in the new solution there is one central APO hosted on the Marketplace. It coordinates the data exchange and the collaborative planning of the participants in the SC.

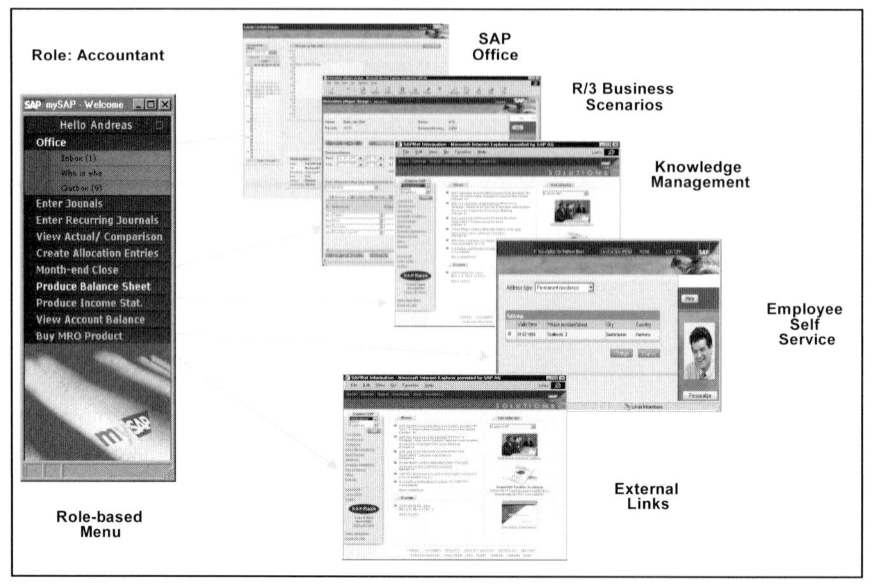

Figure 4.2: Role concept and entry point of the mySAP Workplace

On the next level down in the architecture of the mySAP.com solutions there are several more detailed application systems. Three of them are closely related to SCM.

- mySAP Supply Chain Management

  The core functions of mySAP SCM have been presented in Chapter 3 of this book. The recent developments make intense use of the functionalities provided by the SAP Business Information Warehouse (BW). So there is a powerful functionality for the analysis of data generated by the planning and execution modules of mySAP SCM. It includes the Business Explorer, an analysis and reporting interface with a graphic front-end. Pre-configured data models for business content will reduce the costs for development and maintenance of reports. The BW provides a central entry point to inform the staff of all enterprises in the SC or in the network, allowing them a self-service access to data.

  In spring 2001, more than 2,800 projects were initiated to implement the SAP BW.

- mySAP E-Procurement

  is a partial system close to the EnterpriseBuyer described in section 3.2. It has additional features especially designed for buying via the Internet.

- mySAP Product Lifecycle Management (PLM)

  is the SAP product for PDM, which is described in section 2.1. Participants in modern SC might develop products more and more concurrently. One design goal of mySAP PLM is to prevent a situation where the procedures can only be used by highly specialized engineering teams of large firms. In particular, SME should be able to cooperate in the product development process, also via the Internet.

4.2 Interview with Professor Dr. Claus Heinrich

Claus Heinrich is a member of the board of SAP AG and responsible for the further development of the R/3 System and the SCM products. This interview was conducted in summer 1999.

Mertens: *To be completely consistent, SCM demands central scheduling on the basis of centrally calculated forecasts and centrally applied optimization algorithms. Is this consistent with today's decentralized organizational structures?*

Heinrich: It is not my opinion that planning and scheduling needs to be centralized in the SCM. That would be too complicated. Rather, we need a largely automated synchronization of decentralized planning. The individual departments must have the space to make use of their planning and market knowledge.

Important are fast communication and interaction, but also the willingness to communicate. You can even reward partners in the SC with discounts or bonuses when they improve their forecasts.

Mertens: *Might the philosophy of "Extended enterprise" soon collide with antitrust laws, e.g., when a supply network of involved companies at various production and trading levels is so tightly coordinated that a third party would have to overcome very high entry barriers?*

Heinrich: It is questionable whether the SC will really be so closed. Most involved companies do not want to bind themselves exclusively and for all time to the same partners. The members of the chain are demanding open standards so that they can change their partners too. Although a supplier may well like to be a monopolist for its customers, it would like to choose its own subsuppliers. The number could initially sink, but on the other hand, the Internet technology is lowering the technical entry barriers that run counter to this development.

Mertens: *How can SME "snap in" in future, when for many of them purchasing SAP systems is too expensive?*

Heinrich: The previously mentioned Internet technology, in particular, makes this possible. This is also an ideal example of how technological progress may help SME. Previously they had to make significant investments in order to link to business partners using EDI - nowadays, anyone with a PC and an Internet connection can, for example, take part in a modern delivery schedule.

Mertens: *Can there be an industry-neutral SCM or will industry-specific SCM solutions prevail in the long term?*

Heinrich: We already offer industry-specific solutions. Our APO system is based on central engines and surrounded by industry-specific elements, in the form of a kernel-shell architecture. The core task is, for instance, demand-supply matching. In this regard all SAP SCM systems are similar, and, through the use of the general term resource, we have ensured the necessary abstraction and generality. Industry-specific functionalities and user interfaces are based on this kernel.

Mertens: *We have observed that SAP customers follow quite different goals with SCM and, when viewed more closely, also have very different critical success factors. Is it conceivable that the APO can be customized with parameters for such different situations?*

Heinrich: Customer satisfaction, readiness to deliver, and customer responsiveness are apparently main critical success factors nowadays. However, there are also many other factors that are important for various companies, and they must be able to define these factors individually.

Mertens: *Our research has shown that the APO has not yet attained its original purpose; rather, it is often used as a modern MRP system. Specifically, the APO provides some functionality not present to the same extent in the R/3. Can this mean that the APO will replace the R/3-PP and -MM modules?*

Heinrich: Here I must contradict your proposition. Most companies use APO to optimize their integrated logistics chain. APO, as an extremely powerful production planning system, is an important component, but just one component. APO can be combined with all release levels of R/3 Systems; thus it does not replace PP and MM but rather complements them.

Mertens: *Will SAP transfer the liveCache technology to R/3?*

Heinrich: liveCache is obviously part of the SAP base technology that we use where it provides advantages for our customers. APO can certainly be seen as a pioneering solution with respect to accessing main storage.

Mertens: *Professor Heinrich, you are also an honorary professor at your alma mater, the University of Mannheim. Does APO indicate a renaissance of Operations Research in business management?*

Heinrich: Yes! The computers have now become fast enough to realize the scientific methods developed decades ago.

Mertens: *Your colleague on the board, Prof. Dr. Hasso Plattner, mentioned some time ago the vision of a time when it would be possible to access and process al-*

most all SAP transactions, and not just those in the SCM, via the web. Do you see any such developments?

Heinrich: Definitely. However, today I would no longer call this a vision, but a very specific, short-term development plan. All SAP transactions have been amenable to being run via a web browser since 1999.

Mertens: *It is conceivable that characteristic aggregations of MRP II data and the comparison of demand and capacity at a high aggregation level could be raised to the level of SC networks. In other words: The comparison of sales and production schedules no longer takes place within a company but in the "extended enterprise." Do you consider this to be realistic?*

Heinrich: Certainly! An important advantage of APO is that you can model even beyond MRP II. For example, it is possible to "schedule" in a new supplier company or take account of the limited warehouse capacity at a customer's plant. APO allows you to plan material requirements and to schedule at different levels of companies and supply networks.

5 Case Studies on Supply Chain Management

In this chapter we describe the status of the use of APO in four companies in 1999. We make wide use of material provided by the companies.

5.1 Description

5.1.1 Colgate-Palmolive Case Study

Company profile

Colgate-Palmolive is market leader in the area of dental care and also provides a wide range of shower gels, deodorants and cleaning products. About ten million tubes of Colgate toothpaste are sold worldwide every day.

The company has its headquarters in New York. It has 38,000 employees and is represented in more than 200 countries. The laboratories for research and development are concentrated in Piscataway, N.J. The worldwide turnover in 1999 was approximately $ 9.1 billion with $ 1.5 billion profit.

Objective of the use of APO

The consumer goods industry is struggling with volatile demand resulting from the buying behavior of the retailers. Intensive advertising campaigns cause peaks for the provision of supplies. Further demand peaks result from the buying habits of the end-consumers at the weekend (e.g., bulk buying in supermarkets) and after the payment of wages and salaries.

To master these demand peaks, the retailers set a high value on short lead-times for their orders from the suppliers. These in turn are anxious, in particular in the case of their large customers, to match and integrate the logistics carefully.

Because of the characteristically high standard of cleaning required for the production facilities and the long setup times these entail, the lot size optimization assumes particular importance in this type of consumer goods industry.

In general, the plants should not be set up during a shift. This results in lot sizes being a multiple of the production capacity of a shift, and the lot size planning must be carefully matched with the overall logistics.

Parties involved in the "extended enterprise"

It is obvious that companies will endeavor to use CPFR (cf. section 3.1.2) to solve the aforementioned problems. In this connection, the Colgate-Palmolive group wishes to stock its customers actively at the retailing level. One of its goals is to use VMI (cf. section 2.3.3) to feed 50 percent of the sales volume into the distribution channel in the US. The biggest customer is Wal-Mart with its large number of distribution centers.

Although the stage of a cooperative SCM with the suppliers handled via APO has not yet been achieved, this is planned for the future. Closely related are the so-called co-packers. They add value to the Colgate-Palmolive products by, for example, producing special packaging in the form of vacation or promotion sets.

The independent transport companies do not represent any significant constraints and thus are not included in the APO model.

Structure / network

The Colgate-Palmolive group has 79 factories and distribution warehouses in the US; 280 transport edges are modeled. VMI is practiced with the 109 customer distribution centers also integrated in the supply network, and this number is increasing rapidly. Around 40,000 stock-keeping units are planned with APO. They represent individual articles in their positions, in particular also in warehouses that the supplier maintains in customer plants (Mertens 1995).

As Figure 5.1 illustrates, Colgate-Palmolive is primarily concerned about performing SCM in close cooperation with the retailers and their distribution centers. This "philosophy" is reflected in the organizational structure by the fact that "Customer Account Teams" plan the demand and the deliveries together with the customers. The VMI implementations (cf. section 2.3.3) take account of the heavily fluctuating retail demand. The EDI procedure forms the technological basis using the ANSI X12 exchange format with the message types 852 (Inventory), 830 (Forecast), and 855 (Order).

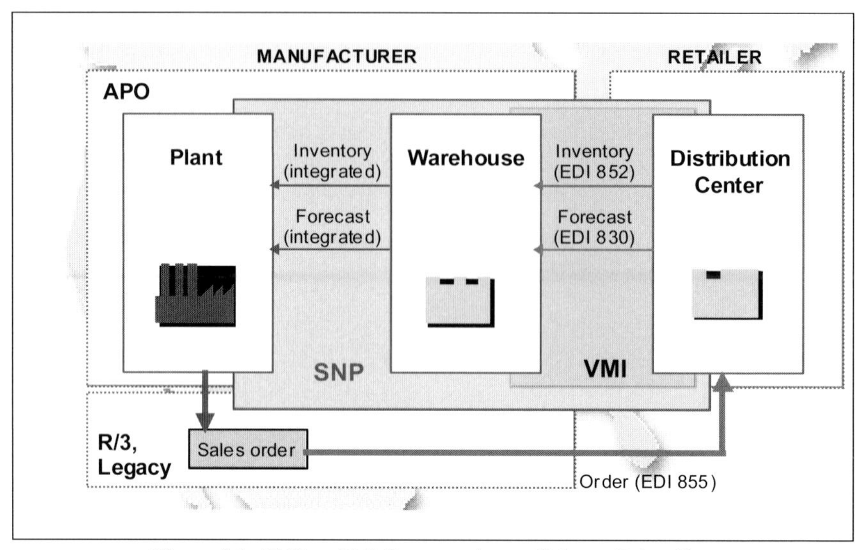

Figure 5.1: SNP and VMI processing at Colgate-Palmolive

IT infrastructure

Colgate-Palmolive has been a user of SAP systems for many years and operates an extensive R/3 installation. One sales region, e.g. Europe or Asia, is assigned to each R/3 System. Each of these systems runs on its own computer, although the computers are all physically installed at the same location (in the Colgate-Palmolive Data Center in New Jersey) (see Figure 5.2). Because the international subsidiary in Mexico enters the CPFR forecasts for its associated warehouse in the parent company's R/3 System, it does not need a dedicated system.

An intensive data exchange takes place, primarily with the customers' distribution centers, either daily or weekly. The incoming data must be checked very carefully. For this reason, data transfer at short time intervals is not considered desirable.

The Internet is currently being used for communication with the suppliers (Supplier Managed Inventory); in future it will also be used in conjunction with the planned CPFR.

APO

Colgate-Palmolive was one of the first large pilot users of the APO. SAP provided special support for the project because even outside the consumer goods industry it is representative for other industries. Executives at Colgate-Palmolive have expressed their satisfaction with the SAP solutions publicly many times.

The group communicated its requirements to SAP in mid-1997. The pilot software was supplied in July 1998. The project team began its work at the same time. APO became productive in April 1999. Table 5.1 describes the project stages.

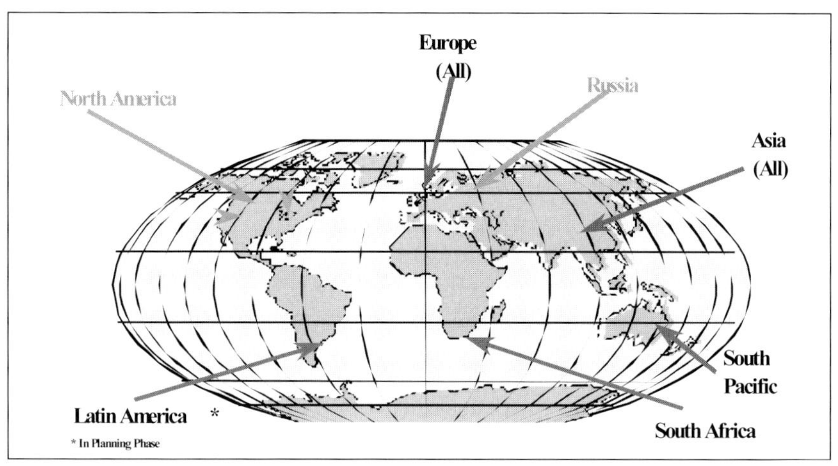

Figure 5.2: Geographic distribution of the IS at Colgate-Palmolive

The Colgate-Palmolive group currently uses APO only in the USA, Canada and Puerto Rico, whereas R/3 is operated worldwide. The following APO modules are used:

- Supply Chain Cockpit, consisting of the SC Engineer for modeling and the SCC for data reports and for management information.
- Demand Planning. This module accepts data from the R/3 Systems implemented at Colgate-Palmolive. Currently, only the forecasting and demand planning functions of the R/3 are used. The data structures for Demand Planning are already stored in the APO. This procedure was adopted to avoid overloading the first project phase.
- SNP: The particular elements used by Colgate-Palmolive are the VMI, DRP, Deployment, and Transport Load Builder subsystems (cf. section 3.1.3).
- PP/DS, in particular with the repetitive manufacturing functionality, will be implemented with the upgrade to Release 3.0 of the APO.

The use of ATP is planned for a later phase. The reason for this, as with the Demand Planning, was to avoid the risk of a "big-bang" conversion.

It is important for Colgate-Palmolive to use ATP, particularly for the so-called backorder processing for customer orders that have already arrived: The customer orders contain priorities that need to be confirmed and acted upon. If necessary, deliveries already confirmed to the customers must be modified.

Stage One
(Parallel Production Ready)
Replaces Manugistics DRP, Deployment and Colgate-Palmolive developed Warehouse Splits, Replenishment Order Prioritizer and Truck Load Building, VMI customer data translator and order generator.
Stage Two
(Implement Remaining VMI Accounts)
Replaces Manugistics DRP, Deployment and Colgate-Palmolive developed, VMI customer data translator and order generator.
Stage Three
(Integrate Canada)
Replaces Colgate-Palmolive developed contingency DRP and deployment for all Canadian-produced products.

Table 5.1: Project stages at Colgate-Palmolive

Colgate-Palmolive plans a long-term coexistence of APO and the R/3 Systems. The latter handle the execution of the plans, e.g., printing of production schedules, delivery notes, and invoices. The close integration within the SAP systems simplifies the reduction of the lead-times for order processing with regard to the goals and success factors mentioned above. The search for a very tight integration of the subsystems resulted in replacement of the Manugistics software previously used for active stocking of the retailer warehouses and planning of the distribution resources. Colgate-Palmolive is also replacing the current transport planning system, which schedules the loading of vehicles and the travel routes, etc., with the corresponding functionality of the APO System.

Use of additional SAP products

Colgate-Palmolive is currently implementing the Business Information Warehouse and is waiting for completion of the Business-to-Business Procurement solution (cf. section 3.2). CRM is planned (cf. section 2.2.4). The use of LES is being considered for transport handling.

5.1.2 fischerwerke Case Study

Company profile

The fischer holding GmbH & Co. KG owns the fischerwerke Artur Fischer GmbH & Co. KG in Tumlingen/Schwarzwald, the Upat GmbH & Co. in Emmendingen/ Baden, and the fischerinternational GmbH, all located in Germany.

The group's business activities are concentrated on fastening technology. fischer is market leader in Europe in the area of tie rods and plugs. It is also active in the fields of automotive systems - storage components for vehicle interiors (e.g., cup holders and spectacle cases) - and model construction sets (fischertechnik).

The number of employees worldwide is almost 3,000. The group owns four factories in Germany and 23 foreign subsidiaries, six of which also produce goods (in Argentina, Brazil, China, Czechia, Italy, Spain). The investment in Germany in 1999 was approximately DM 50 million (≈$ 22 million) - half of which was spent on the new factories in Tumlingen and Denzlingen.

Objective of using of APO

The main objective of using APO, from the viewpoint of the sales and logistics and the associated information processing functions, was to use earlier and more exact forecasts to provide a better basis for scheduling. Previously, a customer, such as a foreign retail chain, ordered its purchasing requirements, which the fischer company responsible outside Germany scheduled on the basis of continuous updates of its inventory. Now, with the goal of Collaborative Forecasting (cf. section 3.1.2), the sales of fischer products to the end-consumers are recorded and a forecast for the complete SC is made from the start. This requires monitoring of many time series, which initially are created decentrally for customers / market regions and summarized for the associated supply warehouses. Even this quantity demands a computerized system. It was also determined at fischer that many stock planners had difficulties in diagnosing the conformity of such time series (e.g., monthly indexes) when they did not have system support. This means that they have sometimes not recognized that an apparent irregularity is the result of a sales promotion. Furthermore, trend changes and trend breakages are recognized too late. This means that major importance must be assigned to the computer-supported reporting of differences between forecast and actual values. Consequently, the forecast methods base provided in the APO promise a significant improvement.

The product managers for production strive to achieve a close working relationship with logistics. They want to overcome not only the barriers between manufacturing and procurement, but also those between production and shipping. Consequently, fischer refers only to "production logistics."

The "construction plan" for the application systems and, in particular, the production planning and control in the fischer group, is characterized by many successively realized solutions. Manufacturing control stations (e.g. the Leitstand FI-2), have long proved themselves and continue to be used in the APO, which is more than just a resource for fischer that supplements the MRP II system in coordination of the SC. Planning is in hand, with a view to the above-mentioned production logistics, for APO to be used to replace both the general planning phases of the MRP II system (capacity planning, material requirements planning) and the detailed production control. In the future, APO will replace the corresponding R/3 modules. The PP/DS module of APO (cf. section 3.1.3) is the one that will be mainly used for this purpose.

A great deal is also expected of the forthcoming simulation functionality of the APO. It should be able to investigate alternatives, for example when larger planned-actual differences arise in a country and inventory must be distributed to other warehouses or when the "feasibility" of a planned production change needs to be analyzed.

Special characteristics of this industry, and thus critical success factors, are:

1. The customers, e.g., craftsmen, are often very late in recognizing what fixings they require. They then need to be supplied promptly to ensure that their work is not delayed. This means short delivery and lead-times for fischer.
2. The degree of automation in the production department is high. It is not unusual for a production order to involve ten hours' setup time and two hours' run time. Thus, improvement of the procedures for lot sizing, capacity scheduling, and utilization rates is of the utmost importance.

Parties involved in the "extended enterprise"

From the procurement viewpoint, fischer's policy is characterized by the use of the fischer APO installation, which makes it possible to give suppliers very early and reliable information about future demands (demand preview). The supplier with the highest priority ("A supplier") should have the option of viewing fischer databases on the Internet, provided this allows it to recognize the demand it needs to meet (but not the total demand!) and thus to optimize its production - in a manner reminiscent of the contract release schedules used in the automotive industry. In return, lower prices are expected from such A suppliers.

Standard parts for which monitoring of quality does not cause any problems (such as packaging material) will also be obtained by way of the SAP Business-to-Business Procurement module (cf. section 3.2) in future, possibly by tendering or even through auctions.

Structure / network

In the logistics chain used hitherto, products were supplied from the central warehouse (Figure 5.3). Currently, 27 warehouse locations exist in central Europe (Figure 5.4).

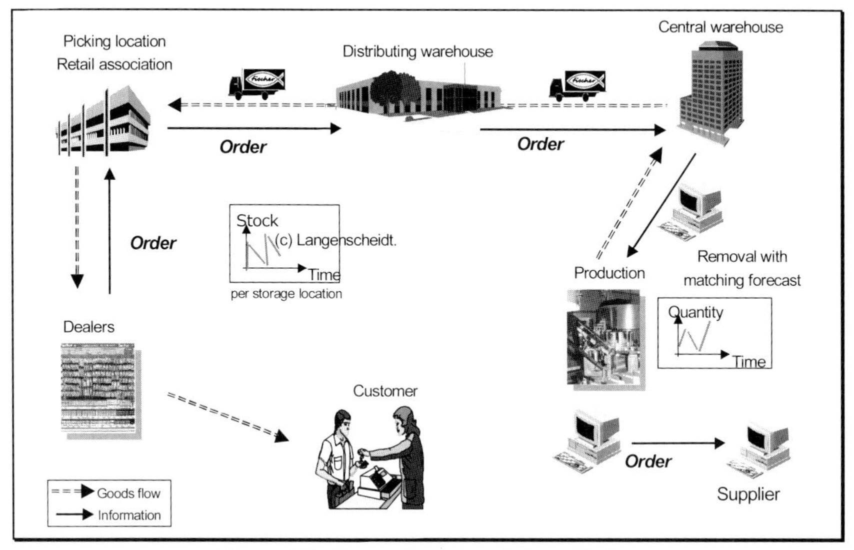

Figure 5.3: Current logistics chain at fischer

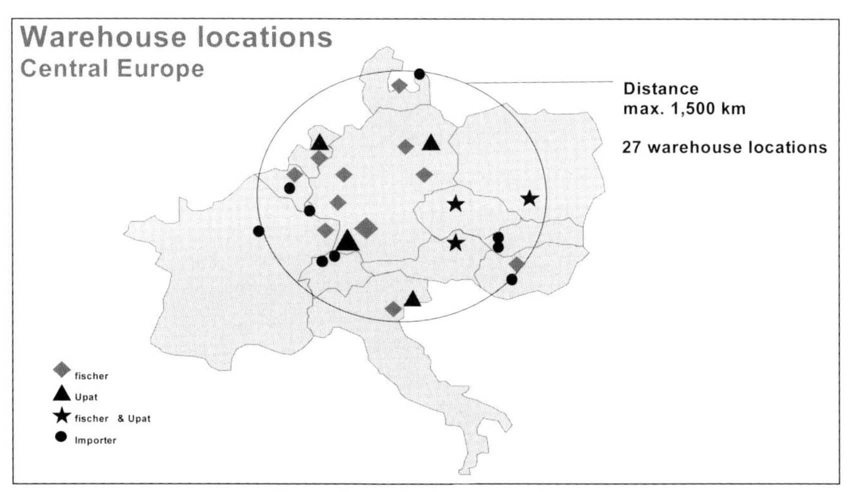

Figure 5.4: Warehouse locations

With the help of the APO, the goal is to achieve the logistics chain shown in Figure 5.5, avoiding the central warehousing facility.

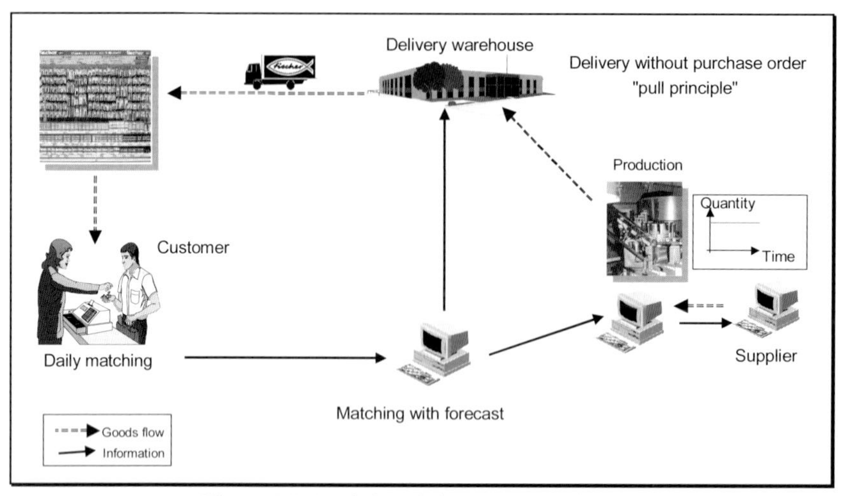

Figure 5.5: Logistics chain planned by fischer

In the medium term, the fischer group will enter into two or three "standard partnerships" with logistics service providers for distribution. Like some suppliers of physical goods, who also receive advance information, these partners will be given advance information on the demand for transport capacity that they can use for their own scheduling.

IT infrastructure

In 1999, fischerwerk was running two R/3 Systems and one R/2 System. The R/2 System is used for sales administration. R/3 4.0b is applied for production, and R/3 3.1i for parts of sales, purchasing, controlling, and inventory management (cf. Figure 5.6).

As hitherto, the information processing will be distributed to the various business divisions ("one APO per division"; see Figure 5.7).

EDI is used in the B2B area, e.g., in communication with car manufacturers and large do-it-yourself retail chains. An increased use of the Internet is planned in the B2C sector, and in the long run also in the relationships between industry and retail companies; both as an ordering medium between retailers and industry, and to allow networking of users (craftsmen), retailers, and suppliers.

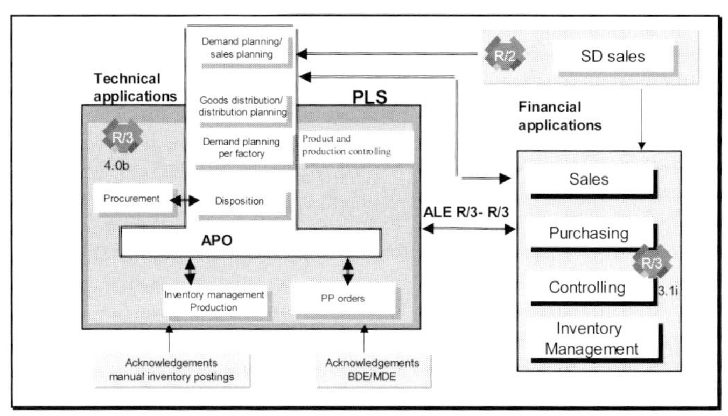

Figure 5.6: SAP R/2, R/3, and APO software architecture (as used at Emmendingen in 1999)

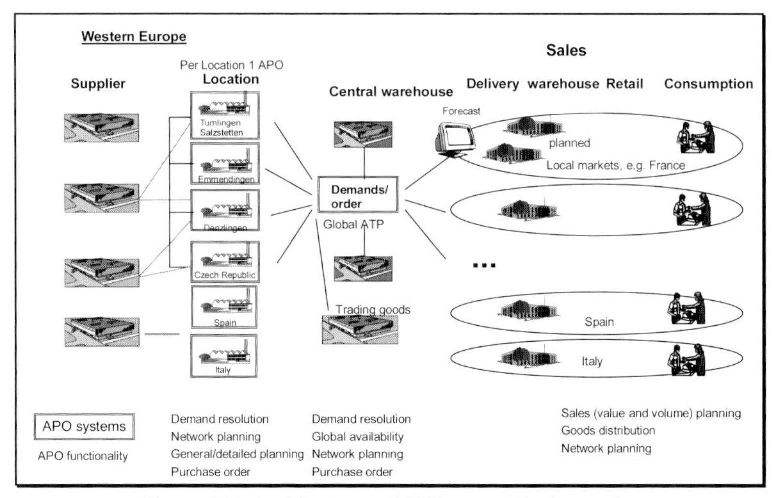

Figure 5.7: Architecture of APO use at fischerwerke

APO

fischerwerke is a pilot customer for the APO. The first APO components have been in use in production in the fischer group since August 1999 (cf. Figure 5.8). Approximately one year elapsed between the initial installation of the system on the fischer hardware platform and its productive use. The total budget for hardware, software (R/3 and APO), and internal costs is approximately $ 2.6 million or DM 6 million.

In the medium term, the fischer group plans to implement all APO modules. The Supply Chain Cockpit is already in use (Figure 5.9). Although the Demand Planning/Forecasting (cf. section 3.1.2) is crucial for meeting critical success factors and the primary goals, PP/DS (cf. section 3.1.3) also has an important role.

The goal is not only an optimal plan, e.g., for machine assignment; rather, self-regulating groups of workers should be informed well in advance of what orders they can expect.

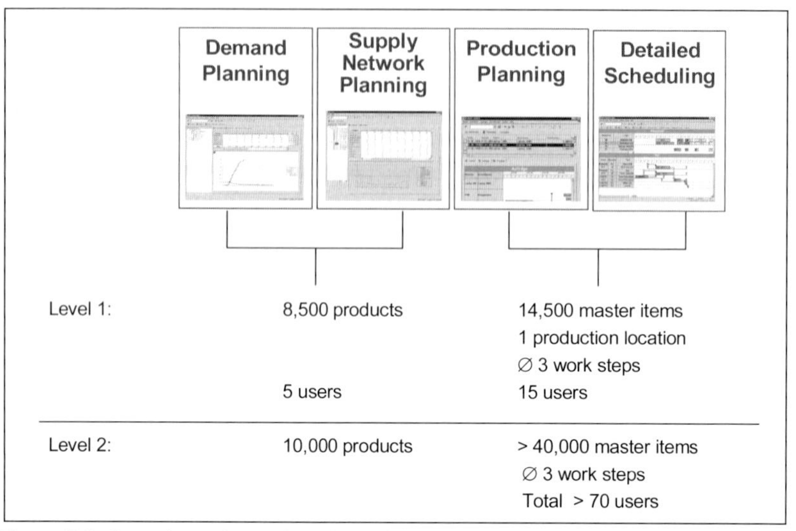

Figure 5.8: Overview of use of APO modules planned in the first and second phases

Because ATP has lost some of its importance at fischer, its use will be considered only in later project phases. The fischer group expects that SAP will make a number of further developments to APO. These include:

1. Increased consideration of the current manufacturing situation (a PDA connection is still lacking)
2. Increased independence of R/3, e.g., for maintaining master data
3. The control of simple orders in made-to-order production in the direction of a project management system
4. Certain workflow management elements
5. Improved coordination with the mobile sales force (cf. section 2.2.4) so that office and field sales personnel can use the same user interface to access the relevant information in the databases
6. Improved representation of the value flow.

Use of additional SAP products

The use of Business Warehouse (BW) is planned and should start on completion of the APO implementation; a BW shall be implemented for each site. fischer also plans "several small Business Warehouses" on the basis of a standardized cube methodology. The use of LES (cf. section 3.3) is not planned at the moment.

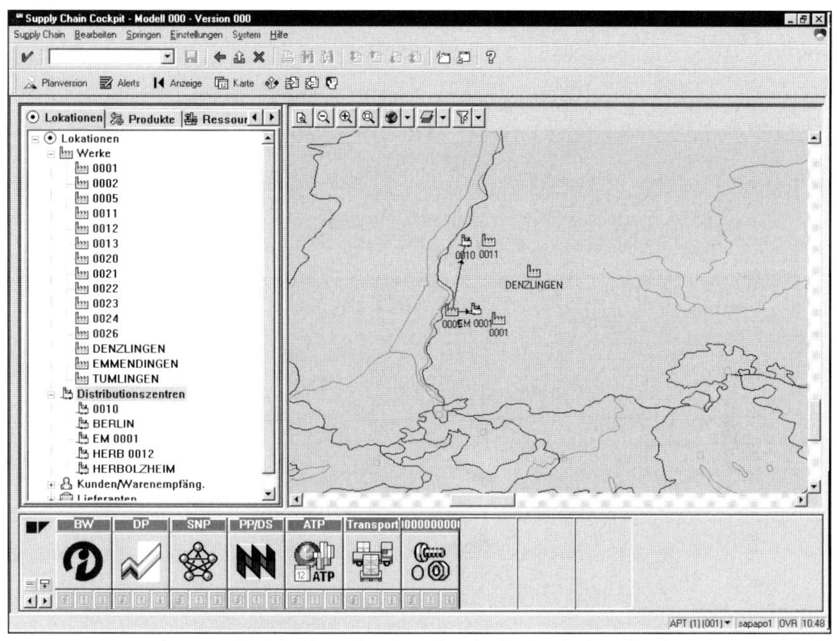

Figure 5.9: A look at the Supply Chain Cockpit at fischer

5.1.3 Salzgitter Case Study

Company profile

Salzgitter AG is the second-largest steel-producing company in Germany. It has three production sites:

1. Peine (profiles)
2. Salzgitter (hot steel / flat steel)
3. Ilsenburg (heavy plate)

The "Peiner Träger" (steel beam), which is known worldwide, is produced at the Peine plant. It is one of the most important elements used in modern steel construction. The manufacture of flat products is concentrated at the Salzgitter plant. Hot wide strip and thin sheet and surface-handled thin sheet with electrolytic or hot-galvanized coating are produced there. Coated thin sheet is one of the most important materials for customers in the automotive manufacturing industry. Salzgitter also has blast-furnace and steel manufacturing metallurgical plants that provide the basis for all flat products and heavy plates.

The Ilsenburg plant in Saxony-Anhalt produces heavy plates from base steel, quality steel and stainless steel, and laminated sheet. These products are used mainly in bridge building and container construction and for manufacturing large pipes.

Salzgitter-Stahlhandel with headquarters in Dusseldorf is a steel trading organization that operates in all important commercial centers of the world. The company is flexible in satisfying the demands of its national and international customers.

In addition to the business areas of steel production and trading, the group is also involved in the area of raw materials and services. The group consists of more than 50 national and international subsidiaries and associated companies. Strategic alliances complement the worldwide activities.

The consolidated group turnover in 1998/99 was DM 5.2 billion (≈$ 2.3 billion). The group has a workforce of more than 12,400 employees, approximately 9,300 of whom work in the plants while approximately 3,100 are employed at the subsidiaries. The basis of the high degree of flexibility is a state-of-the-art production technology. More than DM 850 million (≈$ 370 million) has been invested from the cash flow over the past four years.

Objective of using APO

Even in the steel industry, customers demand very short lead-times in conjunction with JiT supply. Meeting such customer wishes while maintaining quality standards is a critical success factor.

The steel industry is characterized by intensive use of equipment and material. Both result in a significant tie-up of capital. Salzgitter AG uses information processing for optimization, with the goal: "The right material at the right time at the right plant". Stocks held at intermediate warehouses should, then, be minimized and the capacity available used to the fullest possible extent.

A special challenge is to ensure that the production process meets the technological quality characteristics desired by and confirmed to the customer. If there are differences from the specified quality demands, APO should immediately define a new production order or reschedule available items.

Parties involved in the "extended enterprise"

There is a trend for the automotive industry to ask suppliers to deliver complete system components. Many vehicle manufacturers are aiming at future outsourcing costs amounting to 75-80 percent of the total costs.

The customers of Salzgitter AG have planned schedules for their work steps very carefully, e.g., in the body stamping shop. Should the sheet steel supplied from Salzgitter arrive too late, very high additional costs result because of rescheduling and low capacity loading at the customer's site.

Consequently, Salzgitter AG finds it desirable to provide customers with access to their own data. This allows them, for example, to follow up the progress of the orders they have placed. For example, a car manufacturer can see that although Salzgitter AG carries an inventory of steel sheet for motor hoods, it does not carry any stocks of steel sheet for roofs. The vehicle manufacturer thus obtains an additional basis for scheduling its own production. Obviously, this requires secure authorization validation. SAP is planning such a solution in a future release of APO.

Structure / network

The network centers on the three plants in Peine, Salzgitter, and Ilsenburg. The customer is usually in charge of the cooperation with the carriers. The customers provide very short notice for eighty percent of the trucks.

IT infrastructure

The GESIS, Gesellschaft für Informationssysteme mbH, a 100-percent subsidiary of Salzgitter AG, holds 24.5 percent of the shares in the ICG Informationssysteme Consulting und Betriebs-Gesellschaft mbH; the other shareholders are IBM, with 51 percent, and the Continental AG, with 24.5 percent. The ICG operates a computing center that handles large parts of Salzgitter's commercial data processing using an SAP R/2 System. The computing center is located in Salzgitter city. The RV (sales), RF (finance), RM (material), and RK (accounting) components of R/2 are used. In addition, the QSS component developed in-house is used for quality management. QSS is written in the SAP programming language ABAP/4 and integrated with the R/2 System. In-house developments that run independently of the SAP system are used for production planning and material tracking. EDI is used as the interface with the non-SAP systems. Internet applications are planned.

APO

Salzgitter has been a pilot customer for APO since its launch. At Salzgitter's suggestion, SAP added a campaign planning functionality, a typical requirement of the steel industry, to the software package (with the designation "Block Planning").

APO was put into test operation in September 1999 for the planning of the production plants in the Salzgitter works. Once the test operation has been completed successfully, APO will specify to each of the factories which orders are to be processed there and the associated quantities and deadlines.

The following components of the APO package were considered:

- The *Supply Chain Cockpit* is applied only to a limited extent. Only the *Alert Monitor,* an important component of the SCC, is used. It reports bottlenecks and monitors the goods and information streams.
- *Demand Planning* is not used for forecasting.
- *Supply Network Planning (SNP)* is to be implemented in the second project phase.
- *Production Planning and Detailed Scheduling (PP/DS)* will be implemented with all functions for comprehensive detailed planning (see below). The planning will be performed not only at the level of the end products but for all intermediate levels.
- *Available-to-Promise* will be implemented in project phase 2.

Figure 5.10 shows the data flow as planned at the productive start of the APO after the first project phase at Salzgitter AG. The RV module of the SAP R/2 System forms the "gate" to the SCM system.

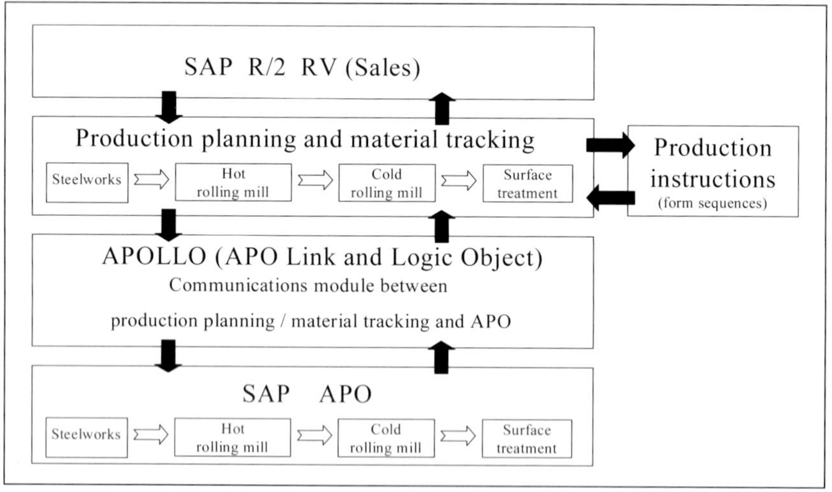

Figure 5.10: APO as incorporated in project phase 1 at Salzgitter AG

A customer order can be split into a maximum of eleven production orders corresponding to the eleven production levels. A capacity unit is assigned to each production level.

The APOLLO (APO, plus Link, Logic, and Object) system developed in-house is used for communication between production planning / material tracking and APO. APOLLO prepares various information, in particular for the material tracking over all production levels. It allows both tracking of individual items and bundling of information. Once the customer orders have been scheduled, the remaining capacity can be obtained and made available for new orders. APOLLO also performs the "translation" of the data into the APO "language" (Figure 5.11).

APO is used primarily to dispatch orders and to determine the schedule; in other words: APO performs the capacity scheduling. APO chooses between alternative resources and takes account of scheduled customer orders, provisional deadlines in production, and shift schedules. The capacities are matched with each other so that an upstream plant always has a larger cross section than the following plant. Excess capacities at the plants are used for external production.

The determination of the production sequences is a complex activity. For example, a direct transition from high-lead to low-lead galvanizing should be avoided. Consequently, all orders with high-lead content must be bundled as a lot, as must all orders with low-lead galvanizing ("campaign building" or, in SAP terminology, "block building"). Thus, a scheduler initially defines fixed campaigns for a longer period. The APO must take account of this and signal any endangered scheduled dates. It is planned that in a later phase the APO will automatically limit the campaign lengths, depending on the order situation.

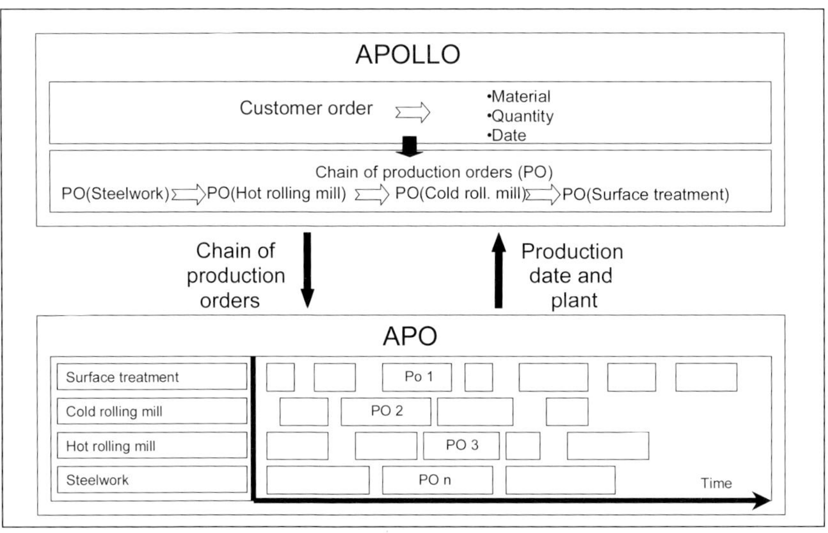

Figure 5.11: Communication between APOLLO and SAP APO

Salzgitter AG plans to implement the following functionality in further project phases:

1. An ATP validation (cf. section 3.1.4)
2. Dynamic Pegging (cf. section 3.1.3), which involves the assignment of items to orders. This assumes that a classification has been made. The same coils must be assigned to the same class (a coil is a semifinished product resulting from hot strip rolling).
3. The SNP module (cf. section 3.1.3) is to be used for including the consignment stores.

After the planning run, APO returns the schedules to APOLLO, which then forwards them to a knowledge-based production assignment system developed in-house. This IS named ESAP (expert system for facility programming [Anlagen-programmierung in German]) and determines detailed sequences.

The feedback from APOLLO allows tracking of individual items. The data are exchanged between the APO and the surrounding systems on a daily basis. A true real-time solution, in which even smaller changes are reported, is currently not needed.

Use of additional SAP products

Implementation of the BW is being considered. It could be used as the basis for replacement of an older reporting system, which gives such information as how many reruns were needed. The use of LES is not planned. Implementation of the Business-to-Business Procurement module may be considered in the long term.

5.1.4 Wacker Siltronic Case Study

Company profile

Wacker Siltronic AG, a 100-percent subsidiary of Wacker Chemie GmbH with its headquarters in Munich, is a leading manufacturer of hyperpure silicon for the semiconductor industry. Together with its parent company, Wacker Siltronic is the only company in the world that produces silicon using a fully integrated manufacturing system. The product spectrum covers ultra-pure silane chloride, polycrystalline pure silicon, crucible- and zone-grown monocrystals, and lapped, etched, polished and epitaxial disks in a wide range of diameters. Silicon from Wacker has been used in the worldwide semiconductor industry since it first began in the early 1950s. Wacker, with its high-tech products, has supported the development of new semiconductor elements from the first silicon transistors through the most advanced current ICs, and onward to future component generations.

Worldwide, around 5,700 employees at factories in Burghausen, Wasserburg and Freiberg, Germany, and Portland, USA, achieved a revenue of approximately \$ 670 million (DM 1.52 billion) in 1999. To provide its global customers with an even better service and to strengthen the position of Wacker Siltronic in the market, an additional factory has been built in Singapore (Figure 5.12).

Objective of the use of APO

Linking of the worldwide semiconductor lines is increasingly gaining strength. All semiconductor manufacturers supplied by Wacker Siltronic with products and services are increasing their global presence. Others are forming alliances in order to use the resulting synergy effect to reduce the total costs of development.

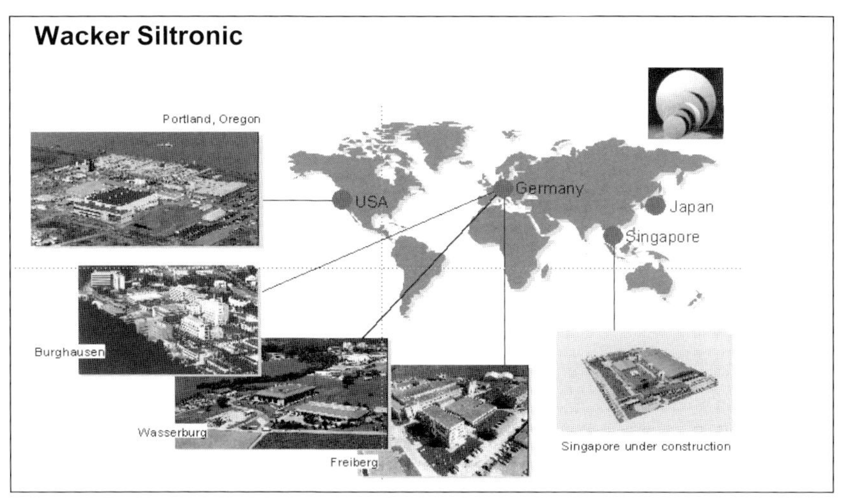

Figure 5.12: Locations of factories

An ambitious "Product to Customer" (PTC) project was started in 1997, aiming at restructuring of all business processes. The emphasis lies on speed, functionality, simplicity, and global integration. The advantages claimed for the PTC project are oriented primarily on the following company goals:

- To be the most important global supplier for the most important customers,
- to provide a complete package that consists of qualitative, high-grade products, technology, information, and service,
- to make the business processes as simple and transparent as possible for the customers, and
- to attain the leading position in profitability for the most important silicon manufacturers.

With the start of the PTC project, the APS systems available on the market were evaluated intensively. Wacker Siltronic gave five main reasons for selecting APO:

1. As a pilot customer, it had the opportunity to influence the development of the software so that no further programming will be needed to represent the processes (principles applied: simplicity, standardization).
2. The integration of existing SAP modules forms the basis for use of all the functions in SAP R/3 (principle applied: functionality).
3. As result of the integration achieved by item 2, the SAP R/3 and APO combination also provides the functional concept for the SCM, because this can ideally combine the logistical with the administrative parts of the processes (principle applied: global integration).
4. The liveCache architecture of APO provides high performance in supporting the business processes (principle applied: speed).
5. It will be possible to adapt very quickly to new releases of the SAP-software (principle applied: use of standard software).

Parties involved in the "extended enterprise"

In the first phase of APO implementation, the immediate customers and suppliers are linked. The integration of further levels on both sides of the SC is possible in a later phase (see Figure 5.13).

The semiconductors division produces for customer requirements in small batches. The industrial customers are supplied directly; no trading companies are used.

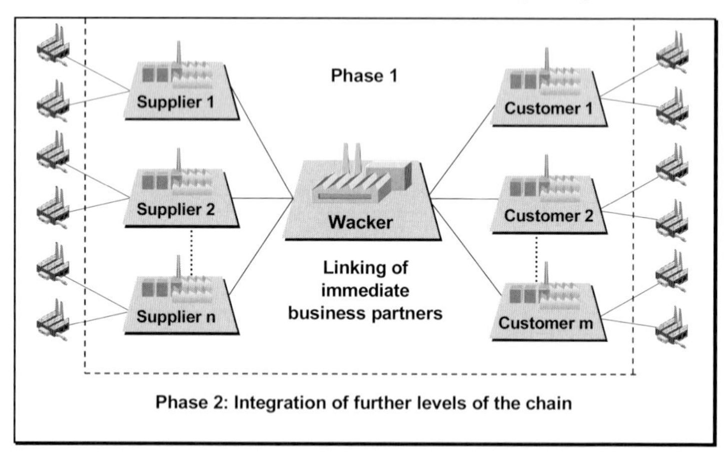

Figure 5.13: Implementation concept of Wacker Siltronic

Structure / network

Orders can be divided between different locations, provided that each is equipped for the associated product.

Cooperation with the carriers is organized differently at the various factory locations. Currently, the company does not see any need to integrate the IS of these service providers into its own IS.

IT infrastructure

The business processes are represented in SAP R/3 and in the APO. This means that the "back-office" functions of SAP R/3 that are already being used productively (modules FI, CO, AA, PM, MM, PS, and HR) are extended by SAP R/3 modules SD, PP, and QM, and by APO.

All locations access the R/3 System running centrally in Burghausen.

EDI is used as the interface with non-SAP systems. Internet connections are not currently used, but their future use is planned.

APO

Wacker has been a pilot customer since development of the APO started. The first installation of the APO has been productive since May 1999 as part of the system installation for the "Product to Customer" business process at Wacker Siltronic's Singapore plant. The installation covers all functions of the planning system that were available with Release 1.1.

Wacker Siltronic considers strategic planning to include both long-term planning based on current market studies and network optimization encompassing all problems defined in "what-if" scenarios.

Long-term planning is handled principally by the Demand Planning module. Interaction of the optimization methods (linear optimization based on "penalty costs") and heuristic methods (Capable to Match as a rule-based planning tool) are exploited within Supply Network Planning to check the production capacities. The network planning algorithm of the Supply Chain Cockpit is used to process "what-if-problems".

The forecast is the most important process in the tactical planning procedures of Wacker Siltronic. The master data maintained in the SAP R/3 are transferred to the APO in an event-controlled manner in the form of characteristic combinations (see Figure 5.14).

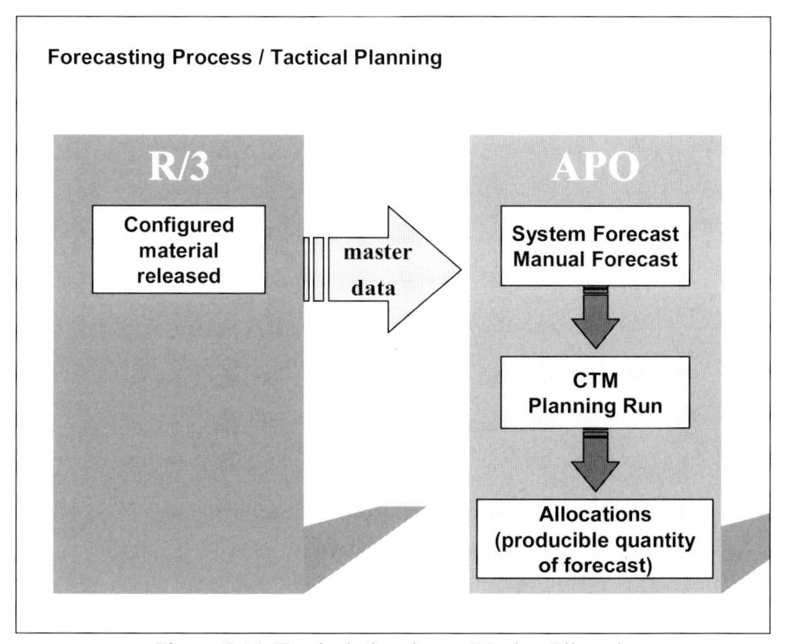

Figure 5.14: Tactical planning at Wacker Siltronic

The forecast values calculated automatically by the Demand Planning module serve as information for the Marketing and Sales Departments. However, the data that have been inputted manually are decisive in the reservation of capacity, which is performed using the Capable-to-Match module. The overall capacity reservation process runs here a background job, returning each result to the corresponding planning book.

Order acquisition is performed basically in two different ways: The traditional method is the acquisition of an order in the SAP R/3 SD module. Using the APO ATP, each order is checked in real-time and in a rule-based manner against the available capacity (taking account of reservations), the inventory, and the work in progress (WIP) for all production levels and for all theoretically possible factory locations.

The second type is that of the so-called SC management orders. The information (forecasts, consumption, inventory) received from the customer is transferred to the APO via the Internet or EDI and determines the quantity of the subsequent delivery. The ATP test is used to create an order in the SAP R/3 and confirm the delivery quantity. At the same time, the system utilizes this calculation to start the production orders so that the next deliveries are satisfied (Figure 5.15).

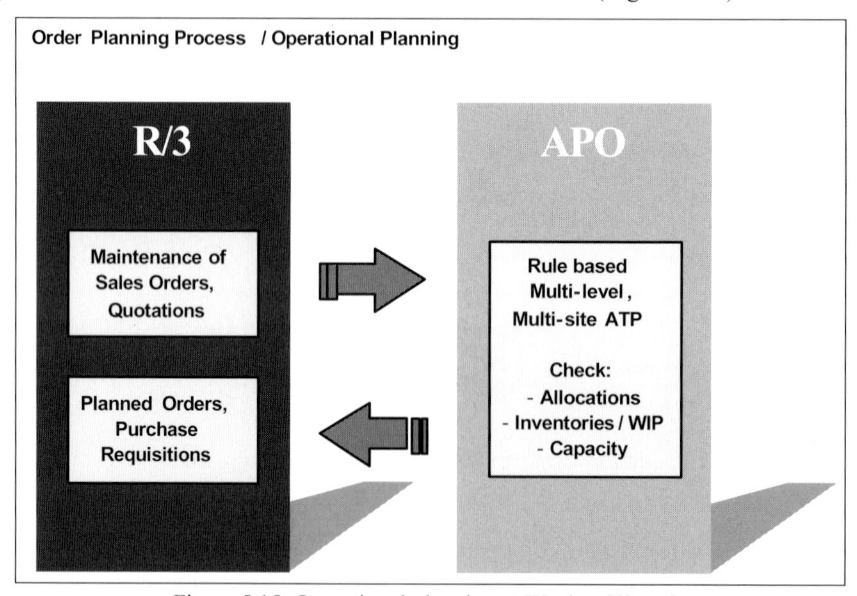

Figure 5.15: Operational planning at Wacker Siltronic

The planning orders created by the APO system (or the PP module) and purchase requisitions are converted in the SAP R/3 into the corresponding production and purchase orders. The production orders are transferred both to the APO and to the "Manufacturing Execution System" (MES). APO uses the appropriate control strategies and optimization profiles to create a plan for each situation and forwards it to the MES using a task list. Confirmations of the production orders are booked automatically both in SAP R/3 and in the APO (Figure 5.16).

Use of additional SAP products

Installation of the Business Warehouse is planned and should start after completion of the installation of the APO. The use of LES and of Business-to-Business Procurement is not planned.

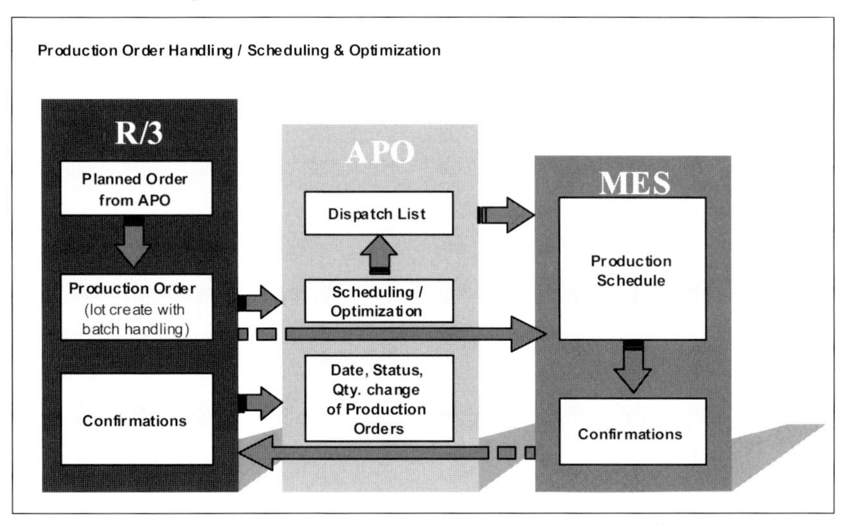

Figure 5.16: Detailed planning and control at Wacker Siltronic

5.2 Summary

Although these four practical examples give an impression of the functionality and usability of the SAP APO package, they are insufficient to allow any broad and general statements.

In our talks with the managers responsible for the SCM implementation in manufacturing companies, we have felt both a demand pull and a technology push. Many specialists have the feeling that systematic agreement in supply networks can bring "rich rewards" in terms of rationalization and improved scheduling. At the same time, a technology push can also be perceived, caused by developments in the area of the Internet and the availability of powerful computer systems that permit the use of the liveCache technology. In this connection, the time seems ripe for a renaissance of mathematical planning and scheduling methods, because, first the computerized aids have become more powerful, secondly, more powerful optimization algorithms have been developed, and, finally, it has been recognized that human planners and schedulers have major problems in maintaining an overall perspective and being able to handle each planning object with the same care when a very large number of interdependent entities are involved. Consequently, it is no surprise that the APO, as a "decision-support methods base" in the area of logistics, has attracted considerable interest in practice. Many practitioners think that the APO does not only fill the gap that has persisted in the scheduling/detailed planning area in many companies - even those with "dense IS architecture". Rather, the APO replaces existing algorithms and already implemented planning and scheduling methods that have not proved entirely satisfactory, such as methods operated as parts of conventional MRP systems.

APO use is currently concentrated very clearly on the reorganization of structures and, in particular, on the processes between plants belonging to the same company or the same group of companies. In other words: There is still little evidence to suggest that "extended enterprises" are really being created. In a four-phase scheme for SCM, in which the first two phases follow internal projects and the two more advanced phases strive to achieve external cooperation, 80% of the companies are currently concentrating on the improvement of internal procedures (Poirier 1999, pp. 23). It is difficult to forecast how fast the development will continue, whether there will be any major setbacks, and whether the SCM concept will be successful in supporting extended enterprises.

6 Internet Resources for Supply Chain Management

SCM is a topic that has attracted wide interest recently, that is to say since the Internet has been widely accessible. Therefore it is not surprising that plenty of information about SCM is available on the web and that the number of SCM-related resources is growing rapidly (Table 6.1).

Table 6.1: Results of searching for the string "supply chain management" on the web

Searching with	Number of hits	
	2000-04-23	2001-04-17
Altavista (Advanced Search)	69,421	111,417
Google	58,200	282,000
Northern Light	86,516	155,298
Yahoo	3 categories; 82,049 web pages	4 categories; 126,000 web pages

The following description of what we regard as the main SCM-websites is based on Knolmayer/Walser (2000); some additional sites are described and changes in the web-sites have been updated. Of course, not all web-addresses and contents will exist in their present form in the future; however, most contents are offered by organizations with a strong focus on SCM, and we assume that they will continue to provide valuable entry points for searches aimed at current information about SCM. Another annotated description of SCM-related online-resources is available in Woods/Marien (2001, pp. 485).

The descriptions of the web-resources are arranged in seven tables; each of them concentrates on a specific type of information providers. If a site-specific search is available, we mention the number of hits obtained when a search for the string "supply chain management" was conducted. These searches were performed at the end of May 2001. However, some of the results may be biased by unreliable or inadequate search engines implemented at some web-sites.

6.1 Portals

Portals are designed as entry-points for web-sessions and often present newsworthy information that is expected to interest all surfers or specific groups of them. Many portals focus on a special community or topic, e.g. within computer science or IS. Table 6.2 provides an overview of SCM-oriented portals. In many cases these portals also contain information about ERP-systems. We recommend the portals About.com and ITtoolbox as entry points.

6.2 Communities, Societies, and Professional Associations

Basic information about a certain topic is often provided on the web by communities, societies, and professional organizations. Some of these use at least parts of the notion SCM in their name (e.g., the Supply Chain Council); many logistics societies also provide worthwhile entry points for searches on SCM (Table 6.3). Links to even more associations are given at

http://www.supply-chain.org/html/sccassociations.cfm

6.3 Research Companies

Several companies that present themselves as "research companies" provide studies on relevant issues in the IS area. However, in most cases this information is only available to users who have signed up to a (usually rather expensive) contract with this company. Those pages that are freely accessible offer rather limited information about SCM (Table 6.4). The boundaries between them and consultancies are sometimes not clearly defined, as many of the "research companies" also offer consulting services.

6.4 Consultancies

Some consultancies have established SCM-consulting as one of their business focuses. Because the web-sites of the multinational consulting companies (in particular the "Big Five") are managed locally, it is rather difficult to identify the

SCM-relevant information, since it can appear on any national server and often no global search is provided (Table 6.5).

6.5 Vendors of SCM Software

Modern SCM is not possible without the use of sophisticated software packages. Sometimes it is difficult to distinguish between vendors of SCM systems and those offering ERP-, Business-to-Business-, CRM-, or EC-solutions. The content of software vendors' web-sites changes very rapidly. Therefore we do not describe their content in detail, but show only suitable entry points for retrieving information on SCM systems; we give the number of hits obtained relating to SCM as an indicator of the quantity of information provided and of the attention that certain packages attract (Table 6.6).

Links to vendors of SCM systems are compiled on the following web-sites:

- Baker Library of the Harvard Business School
  http://www.library.hbs.edu/supplychains.htm
- Business.com: Supply Chain Management,
  Directories "Packaged SCM Solutions" and "Software"
  http://www.business.com/directory/management/
  operations_management/supply_chain_management_scm/
- DMOZ Open Directory Project
  http://dmoz.org/Business/Management/Supply_Chain/
- EBN: The Top Supply-Chain Software Vendors
  http://www.ebnonline.com/story/OEG20010112S0049/
- Logistics World, Directory: "Software"
  http://www.logisticsworld.com/
- LogLink's Supply Chain Sites
  http://www.supplychainsites.com/softwarelinks.asp
- Supply Chain Council
  http://www.supply-chain.org/html/vendors.cfm
- TechRepublic, ERPSuperSite
  http://www.erpsupersite.com/quickfinder.htm
- Yahoo
  http://dir.yahoo.com/business_and_economy/business_to_business/
  computers/software/business_applications/business_management/
  supply_chain_management/.

6.6 Universities

Several universities offer undergraduate, graduate, postgraduate, and executive programs in SCM. Some of the responsible organizational units carry the notion SCM in their names. An overview is presented in Table 6.7.

6.7 Journals

Many journals act either as hybrid or as online journals and present themselves on the web. In Table 6.8 we concentrate on journals that carry the notion "supply chain" in their names. The articles or the information about them are often supplemented by additional information, frequently presented in a similar way to how it would be on a portal. A broad information service is provided by Manufacturing.Net, which dedicates one of its five communities to SC and publishes nine online journals that it regards as relevant for the SC community (Table 6.8).

Table 6.2: SCM-specific portals

Information provider, portal name **Link**	**Content**	**Discussion forum**	**Site-specific search**
About.com Logistics/Supply Chain Site http://logistics.about.com/industry/logistics/cs/supplychainmgmt/	SCM definitions; bibliography and abstracts of more than 10,000 logistics articles; news; newsletter; SCM chatroom every Monday; books via Amazon.com.	Yes. Questions to experts possible.	533 hits; 1,192 hits via bibliography.
CIO-Magazine SCM Research Center http://www.cio.com/forums/scm/	SCM is one of 22 CIO research centers. Articles; links to journals, web-sites, and communities; events; metrics; vendors of ERP systems.	Yes. Questions to experts possible.	Search option for CIO.com and ITworld.com; search of CIO.com provides 1,588 hits.
Createcom.com Supply Chain Management Resource Webpage http://www.createcom.8m.com/	Links to SCM-related associations, publications, software vendors, best practices, market data. Books via Amazon.com.	No.	No search engine available.
ECOMWORLD Portal: Supply Chain Management http://www.ecomworld.com/portals/supplychain/	Articles; whitepapers; news. Books via Amazon.com.	No.	72 hits.
Industry Week's The Value Chain http://www.iwvaluechain.com/	Articles; FAQ; news; events; news; newsletters; benchmarks.	No.	142 hits.
ITradar.com http://itradar.bitpipe.com/data/rlist?t=busofit_10_60_6	Articles from several sources in such categories as SC, SCM, SC automation, SC software, SC integration; some of them accessible free.	Together with TechRepublic.	89 hits.
ITtoolbox Supply Chain http://scm.ittoolbox.com/	Well-structured information: Books; articles; news; newsletters for SCM decision makers and SCM doers; jobs; knowledge database (including software comparisons); access to SCM-relevant portals of software vendors. SCM stock index.	Several topics.	86 hits in SC Directory.

Information provider, portal name **Link**	**Content**	**Discussion forum**	**Site-specific search**
ITworld.com http://www.itworld.com/	8 topics, closest to SCM is "Applications". Request for proposal exchange; news; newsletter; jobs; webcasts.	Several forums.	192 hits.
Logistics World http://www.logisticsworld.com/	SC is one of 18 directories. Glossary; acronyms; detailed directory supporting access to software vendors; vendors of SC software. Three types of memberships allow access to different content.	Logtalk.	184 hits in Logistics Directory.
LogLinks Logistics Links: Transportation and Logistics http://www.loglink.com	Search engine. SC is one of 10 business directories. Structure very detailed, e.g., link directory, business directory, and country profiles; books via Amazon.com.	Logtalk.	184 hits.
LogLinks Supply Chain Sites http://www.supplychainsites.com/	Searchable catalog of web-sites related to SCM, logistics, manufacturing, and transportation.	No.	200 hits (200 seems to be an upper limit).
Manufacturing.Net, Supply Chain Site http://www.manufacturing.net/community/ supplychain/	Events; newsletter; associations; standards; online journals (cf. Table 6.8).	Yes.	165 hits in articles, 25 hits in news.
SCM Competence & Transfer Center http://www.iml.fhg.de/%7Escm-ctc/	Definitions; goals and benefits of SCM-CTC; services; market research; publications; events; 3 categories of links (in German).	No.	No search engine available.
Universität Stuttgart, Fraunhofer Institut für Arbeitswirtschaft und Organisation IAO, SCENE http://www.lis.iao.fhg.de/SCM/ http://www.iao.fhg.de/index_e.hbs	Glossary; publications; news; jobs; vendors of SCM systems; links to associations (in German).	No.	No search engine available.
Supply Chain Today http://www.supplychaintoday.com/	Research; glossary; news; annotated links; templates to format (SCM) cases. MBA programs in SCM.	No.	No search engine available.

Information provider, portal name Link	Content	Discussion forum	Site-specific search
The freight transportation and logistics search site (A subunit of LogLinks) http://www.loggie.com/	SC is one of 19 topics. Search engine for logistics resources, e.g., jobs, software. Books via Amazon.com.	No.	200 hits (200 seems to be an upper limit).
TechRepublic ERPSuperSite http://www.techrepublic.com/ http://www.erpsupersite.com/	Articles; news; events; newsletter (archive starting 2000-07); links to magazines, software vendors, and web-sites; financial data on ERP-vendors. Information on ERP systems via ERP QuickFinder.	Can be initiated. Questions to experts possible.	Via TechRepublic 's Research Index: 145 research documents (fee payable) and papers from 18 companies.
The SupplyChain.com http://www.thesupplychain.com/	Glossary; publications; news; events; archive (starting 2000-01-31); links to online journals.	No.	No search engine available.
Penton Media Total Supplychain.com http://www.totalsupplychain.com/	Archive for several logistics magazines; news; events; links to ERP/SCM resources; search option for products/vendors.	No.	200 hits (200 seems to be an upper limit).

Table 6.3: SCM-related information from communities, societies, and professional organizations

Organization Link	Content	Systematic link collection	Membership	Site-specific search
Montgomery Research Achieving Supply Chain Excellence Through Technology (ASCET) http://www.ascet.com/	Some of the 19 topics covered are "ERP", "Internet-enabled SC", "Next generation SC", and "SC processes". About 150 SCM-related white-papers. Questions to experts possible.	10 categories of links.	Free membership in the ASCET project.	30 hits.
APICS - The Educational Society for Resource Management (formerly known as American Production and Inventory Control Society) http://www.apics.org/	Certification; APICS online bookstore; educational offers; awards; APICS Magazine (summaries of the last 3 issues available); jobs.	6 categories of links.	Different types of membership. Special Interest Group on "Constraints Management".	Search engine announced but not available by May 2001.
Council of Logistics Management (CLM) http://www.clm1.org/	Bibliography and abstracts as via About.com (Table 6.2); books; news; events; information on logistics-related university education; awards; doctoral symposium; mailing list; 14 case studies.	No.	Yes; free full access to website to all readers.	No hits; 23 hits for "supply chain". Via bibliography 1,192 hits.
International Purchasing & Supply Education and Research Association (IPSERA) http://www.ipsera.org/	Category "Information & Communication Technology" with information on ERP and SCM software vendors; links to associations, educational offers, and publications.	12 categories of links.	Yes; limited access to web-site for non-members.	No hits (due to ineffective search engine).
Supply Chain Council http://www.supply-chain.org/ http://www.supply-chain.org/html/scor_overview.cfm	Regional chapters with web-pages in different languages. Supply Chain Operations Reference-model; books; FAQ; news; newsletters (archive starting with 1998-11); jobs.	No.	4 types of membership.	No hits.
The International Society of Logistics (SOLE) (formerly known as Society of Logistics Engineers) http://www.sole.org/	Logistics bibliography; information on advanced education in logistics; jobs; news; events; newsletter (staring 1999-12). Discussion groups for members only.	Many links via "Member Services" -> "Reference Lib."	Five types; some parts of the web-site only accessible for members.	No hits; 1 hit for "supply chain".

Table 6.4: SCM-related information from "research companies"

Research Companies Link	Content	Fees payable	Site-specific search
Aberdeen Group http://www.aberdeen.com/	None of the 22 research areas is SCM, but "E-Business", CRM, and "Enterprise Applications" are covered. Research reports; news; newsletters.	For some contents a fee is payable. Access to "Free Research" after free registration.	1 hit; for "supply and chain and management" 143 hits.
AMR Research http://www.amrresearch.com/ http://www.amrresearch.com/practices/scs.asp	SC layer of AMR Research's E-Business model. Knowledge Center with sections on SC Practice and SAP Practice; news; events; newsletter (after free registration).	Yes. Some information free after registration.	200 hits (200 seems to be an upper limit).
Forrester Research http://www.forrester.com/	None of the 26 coverage areas is SCM, but B2B is covered. Glossary; news; events. TechRankings of software tools.	Yes, otherwise only information on title, author, and release date of documents.	150 hits (150 seems to be an upper limit).
Gartner Group http://www.gartner.com/	None of the 26 focus areas is SCM, but "E-Business", CRM, "enterprise management", and "strategic sourcing" are covered. Vendor selection tools.	Yes, otherwise only few resources accessible.	8 hits; 18 hits for "supply chain".
Giga Information Group http://www.gigaweb.com/	Research papers via eSHOP; events.	Yes. Some information free after registration.	No hits; 14 hits for "supply chain".
IDC http://www.idc.com/	Reports; bulletin; events; newsletter (free access to archives); flash.	Yes. Free access to abstract and date of publication.	190 hits.
Input http://www.input.com/	E-Business advisory program, examining, among other topics, Internet-enabled SCM; news.	Yes, access to search results only after contract.	166 hits in vendor profiles, buyers guides, and reports.
Metagroup http://www.metagroup.com/ http://www.metagroup.com/communities/ccm/ccm.shtml	Metagroup also uses the notion "Commerce Chain Management" for SCM. Research library by topic; news; events; several newsletters; audio documents "Metaview".	Yes, otherwise only few resources accessible.	250 hits (250 is an upper limit in advanced search).
Xephon http://www.xephon.com/	Xephon IS library; journals; reports; news; events; several newsletters.	Yes, but summary and contents of reports free. Free access to some reports.	32 hits.

Table 6.5: SCM-related information from consultancies

Company Link	Content	Methods, tools for SCM	Site-specific search
Accenture http://www.accenture.com/xd/xd.asp?it=enWeb&xd=services/scm/scm_home.xml	9 white papers and reference material; awards; news; events.	Supply Chain Value Assessment; Dynamic Pricing Solutions.	11 hits; more than 200 hits for "supply chain".
CSC http://www.csc.com/solutions/supplychainmanagement/	News; case studies.	C-Discover methodology; Supplier Relationship Management.	245 hits (many redundant)
Deloitte & Touche http://www.us.deloitte.com/mss/improve/speed.htm	Scattered information, e.g., in the section "Management Solutions & Services".	$PEED: Process models, analysis guidelines, SC modeling tools.	100 hits (100 seems to be an upper limit).
Cap Gemini Ernst & Young http://www.cgey.com/solutions/supplychain/	The company also uses the notion "Networked Value Chain" (NVC) for SC. Studies on SCM.	NVC encompasses (e.g.) Information Architecture and Application Integration. Dynamic optimization solutions, agent-based modeling.	53 hits.
IDS Scheer http://www.ids-scheer.com/	News; offerings.	ARIS Collaborative Suite, e.g., ARIS for mySAP.com	45 hits.
IMG - Strategy, Processes, Systems http://www.img.com/english/frameset_e.asp?quelle=e&mo=e&detail=e53	SCM as one of 7 focus areas; news.	eSCM assessment; eSCM optimization; eSCM tool evaluation; eSCM implementation. PROMET methods set.	35 hits.
KPMG http://www.kpmg.com/ http://www.kpmg.com/services/content.asp?l1id=30&l2id=400 http://www.kpmg.ch/mc/scm/home_scm.htm	Scattered information at different country servers; European SC Excellence Award.	e2eSM (enterprise-to-enterprise) SC strategies; Five-stage SC model.	18 hits (at global site).
PriceWaterhouseCoopers http://www.pwcglobal.com/ http://www.pwcglobal.com/extweb/mcs.nsf/$$webpages/supply%20chain%20management	Scattered information.	SAP-related services: ERP systems integration; EC consulting.	896 hits (many redundant).

Table 6.6: SCM-related information from software vendors

SCM software vendors **Link**	**Hits obtained via Altavista (advanced): ("supply chain" OR SCM) NEAR company_name**	**Hits for "supply chain" with local search engine**	**Hits for "supply chain management" with local search engine**
Ariba http://www.ariba.com/	902	225	70
Aspentech http://www.aspentech.com/	183	374	105
Baan http://www.baan.com/	2,232	172	142
Commerce One http://www.commerceone.com/	458	281	52
Descartes Systems Group http://www.descartes.com/	187	47	8
Extricity http://www.extricity.com/	149	73	62
i2 Technologies http://www.i2.com/	4,814	63	52
IBM http://www.ibm.com/services/bustran/supplychain.html	6,636	100 (seems to be an upper limit)	100 (seems to be an upper limit)
ILOG http://www.ilog.com/industries/supplychain/	429	741	114
IMI Industri-Matematik-International http://www.im.se/	164	No local search engine	
J.D. Edwards http://www.jdedwards.com/	2,021	100 (seems to be an upper limit)	3
Logility http://www.logility.com/	430	236	197

SCM software vendors **Link**	**Hits obtained via Altavista (advanced): ("supply chain" OR SCM) NEAR company_name**	**Hits for "supply chain" with local search engine**	**Hits for "supply chain management"with local search engine**
Manhattan Associates http://www.manh.com/	418	19	1
Manugistics http://www.manugistics.com/solutions/scm.asp	2,492	No local search engine	
MAPICS http://www.mapics.com/pointman/pm-supply.html	140	124	89
McHugh Software International http://www.mchugh.com/Solutions/supply_chain/supply_chain.htm	162	125	27
Oracle Supply Chain Management http://www.oracle.com/applications/supplychain/	5,067	200 (seems to be an upper limit)	31
PeopleSoft http://www.peoplesoft.com/en/us/products/applications/scm/	3,524	268	192
Provia Software http://www.provia.com/	67	149	149
QAD http://www.qad.com/product/supply/	505	305 (including irrelevant pages)	220 (including irrelevant pages)
Ross Systems http://www.rossinc.com/manu/manu_scm_ov.html	292	No local search engine	
SAP http://www.sap.com/solutions/scm/	11,232	> 200 (200 is upper limit); (searching the International site)	14 hits; 127 for acronym SCM (sites searched: International, USA, Germany, SAPLabs)
WebMethods http://www.webmethods.com/	352	No local search engine	

Table 6.7: SCM-related information from universities

University Link	Content	Educational level	Links, papers, and additional resources	Site-specific search
Arizona State University (ASU), Tempe College of Business Supply Chain Management http://www.cob.asu.edu/scm/ http://www.cob.asu.edu/mba/day_scm/enews/	SCM FAQ; course descriptions; partner organizations. Newsletter "SCM Links".	B.S. in SCM; SCM specializations in MBA- and PhD- programs. Executive seminars.	5 SCM-links; summary of publications and working papers written at Tempe.	246 hits.
Cardiff Business School, Cardiff Lean Enterprise Research Centre http://www.cf.ac.uk/uwcc/carbs/lerc/	Completed and current research; Supply Chain Development Programme.	Modular MBA in SCM.	Bibliography of publications written in Cardiff.	40 hits.
Cranfield University, Cranfield School of Management Cranfield Centre for Logistics & Transportation (CCLT) http://www.cranfield.ac.uk/som/global_msc/ http://www.logisticsweb.co.uk/	Supply Chain Knowledge Centre (membership required): Library; Journals "International Logistics Abstracts" and "Supply Chain Practice"; LogisticsWeb; e-SC Research Forum; Agile SC Research Club. News. Research projects.	MSc in Logistics and Supply Chain Management.	No.	160 hits.
Ecole Supérieure de Commerce de Bordeaux - Institut Supérieure de Logistique Industrielle (ESC - ISLI) http://www.isli.esc-bordeaux.fr/mh11.cfm http://www.isli.esc-bordeaux.fr/mh31.cfm	Information about program. Editors of "La revue Logistique & Management" (in French; since 1993) and the "Supply Chain Forum: an international journal" (since 2000).	MBA in SCM.	Links to partners in the "SC Network".	No search engine available.
Eidgenössische Technische Hochschule (ETH) Zürich Industrial Management and Manufacturing Engineering http://www.lim.ethz.ch/	Glossary; announcement of books written at ETH Zurich. User Group PIM ("Production and Information Management") (in German).	Lectures and conferences on SCM.	Bibliography of publications written at ETH Zurich; presentation slides; 2 links.	28 hits.

University Link	Content	Educational level	Links, papers, and additional resources	Site-specific search
Harvard Business School (HBS) Supply Chain Community http://www.exed.hbs.edu/program/msc2000/welcome.htm http://www.library.hbs.edu/supplychains.htm	News; articles; bibliography; jobs.	SCM-courses as part of elective curriculum in MBA program.	Yes, via Executive Education.	8 hits in HBS; 14 hits in Working Knowledge; 31 hits in Baker Library (both of the last need registration).
École des HEC, Montreal Group CHAINE http://www.hec.ca/chaine/	Research projects.	Minor SCM in MBA program.	Operations management and logistics links.	43 hits.
Massachusetts Institute of Technology Center of Transportations Studies Integrated Supply Chain Management (ISCM) Program http://web.mit.edu/supplychain/	Research projects; events.	Master Programs in Transportation and Logistics. Executive courses in SCM. Doctoral Program.	Links to SCM-related resources at MIT. Archived newsletters (starting 1995).	618 hits from MIT; 14 hits from ISCM.
Michigan State University, East Lansing The Eli Broad Graduate School of Business Department of Marketing & Supply Chain Management http://www.bus.msu.edu/msc/scm/	Links to members of the SCM Council; newsletters announced but not available (May 2001); event calendar, news about education and information technology, newsletter, partner companies and -organizations; SC benchmarking project.	B.A. in SCM; MS in Logistics; MBA with concentration on SCM; Doctoral Programs in SCM.	4 links.	1,572 hits.

University Link	Content	Educational level	Links, papers, and additional resources	Site-specific search
The Ohio State University, Columbus Fisher College of Business Supply Chain Management Research Group http://fisher.osu.edu/supplychain/ http://www.cob.ohio-state.edu/~logistics/forum.htm	SCM glossary. Survey of career patterns in logistics. Global SC Forum (members only).	Operations & Logistics as major in MBA program.	Well-organized link collection. Articles; Logistics/Distribution manager's bookshelf bibliography (2000-01); annotated SCM bibliography.	502 hits.
Stanford University The Stanford Global Supply Chain Management Forum http://www.stanford.edu/group/scforum/	Events. Partner companies, Global SCM Forum (membership cost is $ 35,000 per year).	SCM is part of masters, PhD, and executive programs.	Summary of case studies. Online access to publications and newsletter "The Supply Chain Connection" (Forum: members only).	10 hits.
Syracuse University Syracuse School of Management http://www.syr.edu/ http://sominfo.syr.edu/depts/supply/	Little information.	SCM as part of undergraduate education. Specialty track SCM in MBA program. SCM is major in PhD program.	3 links to SCM-related sites in Taiwan.	184 hits.
The University of Hong Kong The Poon Kam Kai Institute of Management http://www.pkki.com/cscm3program.html	Little information.	Graduate diploma in e-Supply Chain & Logistics Leadership; Executive certificate in SCM.	Publications originating from Hong Kong.	29 hits.
Universität Bern, Bern Institute of Information Systems http://www.ie.iwi.unibe.ch/	Projects; books (in German).	Lectures in SCM; presentation slides.	Updates of this link list on SCM.	35 hits.

University Link	Content	Educational level	Links, papers, and additional resources	Site-specific search
Universiteit Nyenrode, Breukelen Center for Supply Chain Management (CSCM) http://www.nyenrode.nl/int/research_faculty/cscm/	Description of the Purchasing and Supply Chain Management Unit; research focus.	MBA programs; executive programs; in-company programs; specialist programs.	Not SCM-specific.	23 hits.
University of California, Berkeley Haas School of Business: The UC Berkeley Supply Chain Management Initiative http://haas.berkeley.edu/groups/scmi/ http://haas.berkeley.edu/groups/manit/links.html	Research focus (projects, publications); courses; conferences.	Seminars in SCM; Strategic Supply Chain Management Executive Program.	Yes for Manufacturing and Information Technology. List of publications written in Berkeley.	254 hits.
University of Maryland, College Park Robert H. Smith School of Business Supply Chain Management Center http://www.rhsmith.umd.edu/scmc/	Projects; GLASCO (Graduate Logistics and Supply Chain Organization); FAQ.	MBA concentration in Logistics/SCM. Undergraduate, MBA Logistics/SCM, PhD Program.	Links via GLASCO.	209 hits.
University of Southern California http://www-rcf.usc.edu/~xin/supplychainbookmarks.htm	Little information.	SCM courses as part of undergraduate and MBA programs.	Extensive link collection for an SCM course (as of 1998).	43 hits.
University of Washington John M. Olin School of Business http://www.olin.wustl.edu/classfiles/SUPPLYCHAIN	Guide to SC operations.	SCM as elective course in MBA program.	Links to organizations, consultancies, software vendors, conferences, and online journals.	No search engine available.
University of Wisconsin-Madison School of Business Grainger Center for Supply Chain Management http://wiscinfo.doit.wisc.edu/grainger/	History of the center; newsletter (archive starting 1995-11).	Undergraduate specialization in SCM; MBA and MS in SCM. Executive seminars.	Via Resources grouped links about SCM-related associations, journals, publications, jobs, and other resources.	192 hits.

Table 6.8: Information on SCM journals

Magazines; Publisher Link	Availability of articles, content	Subscription fees	Archive search	Search
European Journal of Purchasing & Supply Management http://www.elsevier.com/inca/publications/store/3/0/4/1/6/index.htt	No.	$ 312.	No.	No.
Purchasing online, The Online Magazine of Total Supply Chain Management and E-Procurement Published by Cahners Business Information http://www.manufacturing.net/magazine/purchasing/	All articles available; software downloads offered; newsletter.	No.	Since 1995.	520 hits.
Supply Chain Forum: an International Journal Published by The Institute of Industrial Logistics of Bordeaux School of Business http://www.supplychain-forum.com/	Published biannually. Some articles available free.	First subscription $ 45.	Since 2000.	No hits.
Supply Chain & Logistics Journal Published by the Canadian Association of Supply Chain & Logistics Management http://www.infochain.org/quarterly/journals.html	Free access for articles published after 1997 and for selected articles published 1996 and 1997.	No.	Since 1996-11.	No.
Supply Chain Management Review Published by Cahners Business Information http://www.manufacturing.net/scl/	About 20 articles and white papers available free.	$ 199 (US); $ 299 (international); $ 99 (educators).	Since 1997.	No.
Supply Chain Management, An International Journal Published by MCB University Press / Emerald http://www.emeraldinsight.com/scm.htm	Title and abstracts accessible free, full text only to subscribers.	$ 649.	Since 1996.	294 hits.
Supply Chain Technology News Published by Penton Media http://subscribe.penton.com/sct/	Weekly newsletter. News; commentary; "best practice" features.	$ 50 in US; $ 70 international; free for qualified.	Since 1999-09.	188 hits.
The Journal of Supply Chain Management: A Global Review of Purchasing and Supply Published by National Association of Purchasing Management http://www.napm.org/Pubs/journalscm/	Abstracts of articles accessible free (starting 2000); titles of articles (1965-2001).	$ 79 (US), $ 89 (international).	Since 1965.	Access to results for members only.

Abbreviations

4PL	→	Fourth Party Logistics
AA	→	Asset Accounting
ABAP	→	Advanced Business Application Programming
ALE	→	Application Link Enabling
AMT	→	Asset Management Tool
API	→	Application Program Interfaces
APICS	→	American Production and Inventory Control Society; Educational Society for Resource Management
APO	→	Advanced Planner and Optimizer
APOLLO	→	APO Link and Logic Object
APS	→	Advanced Planning and Scheduling
ARIS	→	Architecture of Information Systems
ARMS	→	Asset Redeployment Management Systems
ASCET	→	Achieving Supply Chain Excellence through Technologie
ASP	→	Application Service Providing
ATP	→	Available-to-Promise
B2B	→	Business-to-Business
B2C	→	Business-to-Consumer
B2G	→	Business-to-Government
BAPI	→	Business Application Programming Interfaces
BASDA	→	Business Application Software Developers Association
BOM	→	Bill of Materials
BW	→	Business Information Warehouse
C1	→	Commerce One
CAD	→	Computer Aided Design
CALM	→	Canadian Association of Logistics Management
CALS	→	Computer Aided Logistics Support,
	→	Computer Aided Acquisition and Logistics Support,
	→	Continuous Acquisition and Lifecycle Support
CAM	→	Computer Aided Manufacturing
CAPS	→	Center for Advanced Purchasing Studies
CAS	→	Computer Aided Selling
CE	→	Concurrent Engineering
CIC	→	SAP Customer Interaction Center
CIF	→	Core Interface
CIM	→	Computer Integrated Manufacturing
CLM	→	Council of Logistics Management

CLMS → Car Location Message System
CNC → Computerized Numerical Control
CO → SAP Controlling
CP → Constraint Propagation
CPFR → Collaborative Planning, Forecasting and Replenishment
CPU → Central Processing Unit
CRM → Customer Relationship Management
CTI → Computer Telephony Integration
CTM → Capable to Match
cXBL → common XML Business Library
cXML → commercial eXtensible Markup Language
D&B → Dun & Bradstreet
DBM → Database Marketing
DC → Distribution Center
DISRINS → Dismantling and Recycling Information System
DP → Demand Planning
DRP → Distribution Requirements Planning
DS → Detailed Scheduling
EC → Electronic Commerce
ECR → Efficient Consumer Response
Ed., Eds. → Editor(s)
EDI → Electronic Data Interchange
EH&S → SAP Environmental, Health & Safety
ERP → Enterprise Resource Planning
ESAP → Expert System for Facility Programming (in German: Anlagenprogrammierung)
FI → SAP Finance
FMI → Food Marketing Institute
FTL → Full Truckload
GA → Genetic Algorithms
GE → General Electric
GM → General Motors
GPS → Global Positioning System
HR → SAP Human Resources
HTML → HyperText Markup Language
HU → Handling Units
IAC → Internet Application Components
ICT → Information & Communication Technology
IDIS → International Dismantling Information System
IDOCs → Intermediate Documents
IGES → Initial Graphics Exchange Specification
InfoCubes → Information Cubes
IPPE → Integrated Product and Process Engineering
IPSERA → International Purchasing & Supply Education and Research Association
IRI → Information Resources, Inc.
IS → Information System
ISCM → Integrated Supply Chain Management
IT → Information Technology
ITS → SAP Internet Transaction Server
JiT → Just-in-Time
LASeR → Life-cycle Assembly Serviceability and Recycling Prototype Program

LES	→	SAP Logistics Execution System
LIS	→	SAP Logistics Information System
LTL	→	Less Than Truckload
MASIF	→	Mobile Agent System Interoperability Facility
MES	→	Manufacturing Execution System
MIT	→	Massachusetts Institute of Technology
MM	→	SAP Material Management
MRO	→	Maintenance, Repair and Operations
MRP	→	Material Requirements Planning
NC	→	Numerical Control
n.d.	→	no date
NECO	→	Navy Electronic Commerce Online
NIMIS	→	Networked Inventory Management Systems
OAGIS	→	Open Applications Group Integration Specifications
OASIS	→	Organization for the Advancement of Structured Information Standards
OBI	→	Open Buying on the Internet
OLAP	→	Online Analytical Processing
OLTP	→	Online Transaction Processing
OMG	→	Object Management Group
OPL	→	Optimization Programming Language
OPT	→	Optimized Production Technology
ORM	→	Operating Resource Management
ORMS	→	Operating Resource Management System
PC^2	→	Product Change Collaboration
PDA	→	Production Data Acquisition
PDM	→	Product Data Management
PLM	→	Product Lifecycle Management
PM	→	SAP Plant Maintenance
POI	→	Production Optimization Interface
POS	→	Point of Sale
PP	→	SAP Production Planning
PP/DS	→	Production Planning / Detailed Scheduling
PPM	→	Production Process Model
PRPC	→	Production and Recycling Planning and Control
PS	→	SAP Project System
PTC	→	Product to Customer
PVS	→	Product Variant Structure
PWC	→	PriceWaterhouseCoopers
QM	→	SAP Quality Management
REA	→	SAP Recycling Administration
RF	→	Radio Frequency
RFC	→	Remote Function Call
ROP	→	Reorder Points
RV	→	SAP R/2 Sales (in German: Vertrieb)
SC	→	Supply Chain
SCC	→	Supply Chain Cockpit
SCE	→	Supply Chain Execution
SCM	→	Supply Chain Management
SCOPE	→	Supply Chain Optimization, Planning and Execution
SCOR	→	Supply Chain Operations Reference Model
SCOREx	→	Supply Chain Optimization and Real-Time Extended Execution

SCP	→	Supply Chain Planning
SCS	→	Supply Chain Solutions
SD	→	SAP Sales and Distribution
SD-CAS	→	SAP Sales and Distribution-Computer Aided Selling
SE	→	Simultaneous Engineering
SET	→	Secure Electronic Transaction
SM	→	Service Management
SME	→	Small and Medium sized Enterprises
SND	→	Supply Network Design
SNP	→	Supply Network Planning
SOLE	→	Society of Logistics
STEP	→	Standard for the Exchange of Product Model Data
TLB	→	Transport Load Builder
TMS	→	SAP Transport Management System
TOC	→	Theory of Constraints
TP	→	Transportation Planning
TPN	→	Trading Process Network
TP/VS	→	Transportation Planning / Vehicle Scheduling
UN/SPSC	→	United Nations / Standard Products & Services Classification
UTC	→	United Technologies Corporation
VAN	→	Value Added Network
VDA-FS	→	Verband der Automobilindustrie – Flächenschnittstelle (Association of the German Automotive Industry - surface data interface)
VDI	→	Verein Deutscher Ingenieure (Association of German Engineers)
VICS	→	Voluntary Interindustry Commerce Standards
VMI	→	Vendor Managed Inventory
VPDM	→	Virtual Product Data Management
WIP	→	Work in Progress
WMS	→	SAP Warehouse Management System
WWW	→	World Wide Web
XML	→	eXtensible Markup Language
XRP	→	EXtended Enterprise Resource Planning

References

Adelsberger, H.H., Kanet, J.J. (1991) The "LEITSTAND" - A new tool for computer-integrated manufacturing. Production and Inventory Management Journal 32/1, pp. 43-48

Agile Software (1999) Product Change Collaboration in a supply chain centric world. http://www.agilesoft.com/wp/wppc2.pdf[as of 2001-05-15]

Alabas, C., Altiparmak, F., Dengiz, B. (2000) The optimization of number of kanbans with genetic algorithms, simulated annealing and tabu search. In: IEEE Neural Networks Council (Ed.) Proceedings of the 2000 Congress on Evolutionary Computation. IEEE, Piscataway, pp. 580-585

Alpar, P. (1998) Kommerzielle Nutzung des Internet. 2nd ed., Springer, Berlin et al.

Alt, R., Grünauer, K.M., Reichmayr, C., Zurmühlen, R. (2000) Electronic commerce and supply chain management at 'The Swatch Group'. In: Österle, H., Fleisch, E., Alt, R. (Eds.) Business Networking, Shaping Enterprise Relationships on the Internet. Springer, Berlin et al., pp. 127-142

Alvord, C.H. (1999) The S in APS, in: IIE Solutions 31/10, pp. 38-41

Anderson Jr., E.G., Fine, C.H., Parker, G.G. (2000) Upstream volatility in the supply chain: The machine tool industry as a case study. Production and Operations Management 9/3, pp. 239-261

Angeli, A., Streit, U., Gonfalonieri, R. (2001) The SAP R/3 guide to EDI and interfaces. 2nd ed., Braunschweig et al., Vieweg

Arntzen, B.C., Brown, G.G., Harrison, T.P., Trafton, L.L. (1995) Global supply chain management at Digital Equipment Corporation. Interfaces 25/1, pp. 69-93

Ashford, M. (1997) Developing European logistics strategy. Case 5: NIKE Europe. In: Taylor, D. (Ed.) Global Cases in Logistics and Supply Chain Management. International Thomson Business Press, London et al., pp. 61-71

Aviv, Y., Federgruen, A. (1998) The operational benefits of information sharing and Vendor Managed Inventory (VMI) programs. http://www.olin.wustl.edu/faculty/aviv/papers/vmr.pdf [1998-03; 1999-06-09; as of 2001-05-29; under revision]

Baganha, M.P., Cohen, M.A. (1998) The stabilizing effect of inventory in supply chains. Operations Research 46/3 Suppl. Issue, pp. S72-S83

Barratt, M., Ellis, H., Bytheway, A. (1998) Supply and logistics networks - Rethinking the supply chain: Complexity theory and the role of information. In: Pelton, L.E., Schnedlitz, P. (Eds.) Marketing Exchange Colloquium. American Marketing Association, Vienna, pp. 93-106

Barták, R. (1999) Constraint programming: In pursuit of the holy grail, in: Proceedings of WDS99, Prague
http://kti.ms.mff.cuni.cz/~bartak/downloads/WDS99.pdf [1999-12-06; as of 2001-05-15]

Bartsch, H., Bickenbach, P. (2001) Supply Chain Management mit SAP APO. 2nd ed., Galileo Press, Bonn

Bartsch, H., Teufel, T. (2000) Supply Chain Management mit SAP APO. Galileo Press, Bonn

Basu, K., Chandra, P. (1996) Market driven manufacturing: A requirement analysis. Vikalpa - The Journal of Decision Makers 21/4, pp. 19-34

Baumgaertel, H., Brueckner, S., Parunak, V., Vanderbok, R., Wilke, J. (2001) Agent models of supply network dynamics. In: Billington, C., Harrison, T., Lee, H., Neale, J. (Eds.) The Practice of Supply Chain Management. Kluwer, Boston et al.
http://www.anteaters.net/~sbrueckner/publications/2001/AgentModelsOfSupplyNetworkDynamics_submitted_20010209.pdf
[2001-02-09; as of 2001-05-15]

Baumgarten, H., Wolff, S. (1999) The next wave of logistics: Global supply chain e-fficiency. TU Berlin, Berlin/Boston

Baumgartner, H. (1991) Kriterien für eine optimale Beschaffungslogistik. In: Rupper, P. (Ed.) Unternehmenslogistik. 3rd ed., Industrielle Organisation/TÜV Rheinland, Zürich, pp. 63-82

Baumhardt, J.J. (1986) The effect of complexity on product costs and profits. Cause and effect costing. In: Wildemann, H. et al. (Eds.) Strategische Investitionsplanung für neue Technologien in der Produktion. Tagungsband Teil 2. gfmt, Passau, pp. 779-810

Becker, J., Rosemann, M. (1993) Logistik und CIM - Die effiziente Material- und Informationsflußgestaltung im Industrieunternehmen. Springer, Berlin et al.

Bjorksten, A., Knopf, M., Briesch, R. (1999) Enterprise software: The front and back office: Overview and emerging trends.
http://cci.bus.utexas.edu/research/white/ent_soft.htm [1999-05; 2001-05-01; as of 2001-05-15]

Blackburn, J.D. (1991) The quick-response movement in the apparel industry: A case study in time-compressing supply chains. In: Blackburn, J.D. (Ed.) Time-Based Competition: The Next Battleground in American Manufacturing. Irwin, Homewood, pp. 246-269

Blackwell, R.D. (1997) From mind to market. Reinventing the retail supply chain. Harper Business, New York

Bockmayr, A., Kasper, T. (1998) Branch and infer: A unifying framework for integer and finite domain constraint programming. INFORMS Journal on Computing 10/3, pp. 287-300

Bothe, M. (1999) Supply Chain Management mit SAP APO - Erste Projekterfahrungen. HMD - Praxis der Wirtschaftsinformatik 36/207, pp. 70-77

Bourke, R.W. (1997) PDM and ERP continuing to converge. APICS Online Edition 7/8. http://www.apics.org/magazine/aug97/bourke.htm [1997-08; as of 2000-08-09]

Boutellier, R., Locker, A. (1998) Beschaffungslogistik. Mit praxiserprobten Konzepten zum Erfolg. Hanser, München et al.

Bovet, D., Martha, J. (2000) Value nets: Breaking the supply chain to unlock hidden profits. Wiley, New York et al.

Bozdogan, K., Deyst, J., Hoult, D., Lucas, M. (1998) Architectural innovation in product development through early supplier integration. R&D Management 28/3, pp. 163-173

Bramel, J., Simchi-Levi, D. (1997) The logic of logistics. Theory, algorithms, and applications for logistics management. Springer, New York et al.

Brander, A. (1995) Forecasting and customer service management. Optimizing the supply chain with a practical approach. Helbing & Lichtenhahn, Basel et al.

Brockhoff, K. (1979) Delphi-Prognosen im Computer-Dialog. Mohr, Tübingen

Brody, A.B., Gottsman, E.J. (1999) Pocket BargainFinder: A handheld device for augmented commerce. In: Proceedings of the First International Symposium on Handheld and Ubiquitous Computing. Springer, Berlin et al., pp. 44-51

Bullinger, H.-J., Hilty, L.M., Rautenstrauch, C., Rey, U., Weller, A. (Eds.) (1998) Betriebliche Umweltinformationssysteme in Produktion und Logistik. Metropolis, Marburg

Burbidge, J.L. (1962) The principles of production control. Macdonalds/Evans, Estover

Bussiek, T., Stotz, A. (1999) Optimierung der Extended Supply Chain mittels Internet-Lösungen für die Beschaffung (SAP Business-to-Business Procurement). HMD - Praxis der Wirtschaftsinformatik 36/207, pp. 35-46

Buxmann, P., König, W. (2000) Inter-organizational cooperation with SAP systems, Springer, Berlin et al.

Cachon, G., Fisher, M. (2000) Supply chain inventory management and the value of shared information. Management Science 46/8, pp. 1032-1048

Cachon, G.P., Lariviere, M.A. (2001) Contracting to assure supply: How to share demand forecasts in a supply chain. Management Science 47/5, 629-646

Camm, J.D., Chorman, T.E., Dill, F.A., Evans, J.R., Sweeney, D.J., Wegryn, G.W. (1997) Blending OR/MS, judgment, and GIS: Restructuring P&G's supply chain. Interfaces 27/1, pp. 128-142

Carlsson, C., Fullér, R. (2001) Reducing the bullwhip effect by means of intelligent, soft computing methods. Proceedings of the 34th Hawaii International Conference on Systems Sciences. IEEE, Los Alamitos

Carr, C., Ng, J. (1995) Total cost control: Nissan and its U.K. supplier partnerships. Management Accounting Research 6/4, pp. 347-365

Carroll, B. (2000) Guide to selecting a web-based PDM system. http://www.pdmic.com/presenting/webPDM/ [2000-06-19; as of 2001-05-15]

Casey, R. (1996): CALS. http://arri.uta.edu/ccc/calsover.html [1996-03-09; as of 2001-05-15]

Chandra, P., Fisher, M.L. (1994) Coordination of production and distribution planning. European Journal of Operational Research 72/3, pp. 503-517

Chen, F. (1999) Decentralized supply chains subject to information delays. Management Science 45/8, pp. 1076-1090

Chow, W.M. (1965) Adaptive control of the exponential smoothing constant. The Journal of Industrial Engineering 16/5, pp. 314-317

Christopher, M. (1998) Logistics and supply chain management. Strategies for reducing cost and improving service. 2nd ed., Financial Times/Pitman, London et al.

CIMdata (1998) Product Data Management: The definition. An introduction to concepts, benefits, and terminology. http://www.cimdata.com/homepage/downloads/USTECH.pdf [1998-12-11; as of 1999-07-31]

Clemons, E.K., Reddi, S.P., Row, M.C. (1993) The impact of information technology on the organization of economic activity: The "move to the middle" hypothesis. Journal of Management Information Systems 10/2, pp. 9-36

Cohen, S.S., Zysman, J. (1987) Manufacturing matters. Basic Books, New York

Committee on Supply Chain Integration, Board on Manufacturing and Engineering Design, Commission on Engineering and Technical Systems, National Research Council (2000) Surviving supply chain integration: Strategies for small manufacturers. National Academy Press, Washington

Cooper, M.C., Lambert, D.M., Pagh, J.D. (1997) Supply chain management: More than a new name for logistics. International Journal of Logistics Management 8/1, pp. 1-14

Cooper, R., Slagmulder, R. (1999) Supply chain development for the lean enterprise, Interorganizational cost management. The IMA Foundation for Applied Research/Productivity, Montvale/Portland

Copacino, W.C. (1997) Supply chain management. The basics and beyond. St. Lucie Press/ APICS, Boca Raton et al.

Corsten, H., Gössinger, R. (Eds.) (1998) Dezentrale Produktionsplanungs- und -steuerungs-Systeme. Kohlhammer, Berlin

Cox, J.F., Spencer, M.S. (1998) The constraints management handbook. St. Lucie Press/ APICS, Boca Raton/Falls Church

CPFR (1998): Collaborative planning forecasting and replenishment voluntary guidelines, Draft, Internal Paper

Croom, S., Romano, P., Giannakis, M. (2000) Supply chain management: an analytical framework for critical literature review. European Journal of Purchasing & Supply Management 6/1, pp. 67-83

Croston, J.D. (1972) Forecasting and stock control for intermittent demand. Operational Research Quarterly 23/3, pp. 289-303

Daganzo, C.F. (1999) Logistics systems analysis. 3rd ed., Springer, Berlin et al.

Dasgupta, P., Narasimhan, N., Moser, L.E., Melliar, Smith, P.M. (1999) MAgNET: Mobile Agents for Networked Electronic Trading. http://beta.ece.ucsb.edu/~pdg/research/papers/MAgNEThtml/ [1999-01-28; as of 2001-05-15]

Daugherty, P.J. (1994) Strategic information linkage. In: Robeson, J.F., Copacino, W.C. (Eds.) The Logistics Handbook. Free Press, New York et al.

Deloitte Consulting (1999a) Energizing the supply chain. Trends and issues in Supply Chain Management. Toronto. http://www.dc.com/obx/script.php?Name=getFile&reportname=energizing_the_supply_chain.pdf&type=pdf [1999-03-26; as of 2001-05-15]

Deloitte Consulting (1999b) Only two percent of manufacturers rate their supply chain world class. http://www.dc.com/whatsnew/supply_chain.html [1999-03-29; as of 1999-07-31]

Dempster, M.A.H., Fisher, M.L., Jansen, L., Lagweg, B.J., Lenstra, J.K., Rinnooy Khan, A.H.G. (1981): Analytical evaluation of hierarchical planning systems. Operations Research 29/4, pp. 707-716

Dettmer, H.W. (1997) Goldratt's Theory of Constraints. A systems approach to continuous improvement. ASQ Quality Press, Milwaukee

Dittrich, J., Mertens, P., Hau, M. (2000) Dispositionsparameter von SAP R/3-PP. 2nd ed., Vieweg, Braunschweig et al.

Dokaupil, E. (1999) Get together. Erste Studie über SCM bei R/3-Anwendern. Efficient Extended Enterprise no vol./07-08, pp. 18-21

Dolmetsch, R., Fleisch, E., Österle, H. (2000) Electronic commerce in the procurement of indirect goods. In: Österle, H., Fleisch, E., Alt, R. (Eds.) Business Networking, Shaping Enterprise Relationships on the Internet. Springer, Berlin et al., pp. 201-217

Dorn, J., Kerr, R., Thalhammer, G. (1998) Maintaining robust schedules by fuzzy reasoning. In: Drexl, A., Kimms, A. (Eds.) Beyond Manufacturing Resource Planning (MRP II). Springer, Berlin et al., pp. 279-306

Dornier, P.-P., Ernst, R., Fender, M., Kouvelis, P. (1998) Global operations and logistics. Wiley, New York et al.

Drexl, A., Fleischmann, B., Günther, H.-O., Stadtler, H., Tempelmeier, H. (1994) Konzeptionelle Grundlagen kapazitätsorientierter PPS-Systeme. Zeitschrift für betriebswirtschaftliche Forschung 46/12, pp. 1022-1045

Dyer, J.H. (1996) Specialized supplier networks as a source of competitive advantage: Evidence from the auto industry. Strategic Management Journal 17/4, pp. 271-291

Eddigehausen, W. (2000): Understanding SAP APO, gain the knowledge to implement APO. No location, no publishing house (ISBN 0-615-11500-4)

Eisert, U., Geiger, K., Hartmann, G., Ruf, H., Schindewolf, S., Schmidt, U. (2000) mySAP Product Lifecycle Management: Strategie, Technologie, Implementierung. Galileo Press, Bonn

Ellram, L.M. (1999) The role of supply management in target costing. http://www.capsresearch.org/ReportPDFs/TargetCostingAll.pdf [1999-01; 2000-10-27; as of 2001-05-15]

Environmental Protection Agency (2000a) The lean and green supply chain: A practical guide for material managers and supply chain managers to reduce costs and improve environmental performance. EPA 742-R-00-001.
http://www.epa.gov/oppt/acctg/Lean.pdf [2000-01; as of 2001-05-15]

Environmental Protection Agency (2000b) Enhancing supply chain performance with environmental cost information: Examples from Commonwealth Edison, Andersen Corporation, and Ashland Chemicals. EPA 742-R-00-002.
http://www.epa.gov/oppt/acctg/eacasestudies.pdf [2000-04; as of 2001-05-15]

Evers, J., Oerlemans, B. (2001) Ariba to end collaboration with i2. IDG News Service http://www.itworld.com/App/441/IDG010510aribai2/ [2001-05-10; as of 2001-05-15]

Faber, A. (1998) Global Sourcing. Möglichkeiten einer produktionssynchronen Beschaffung vor dem Hintergrund neuer Kommunikationstechnologien. Lang, Frankfurt am Main et al.

Fernández-Rañada, M., Gurrola-Gal, F.X., López-Tello, E. (1999) 3C - A proven alternative to MRPII for optimizing supply chain performance. St. Lucie Press, Boca Raton et al.

Fine, C.H. (1998) Clockspeed. Winning industry control in the age of temporary advantage. Perseus Books, Reading

Fine, C.H. (2000) Clockspeed-based strategies for supply chain design. Production and Operations Management 9/3, pp. 213-221

Fisher, M., Hammond, J., Obermeyer, W., Raman, A. (1994) Making supply meet demand in an uncertain world. Harvard Business Review 74/3, pp. 83-92

FMI (2000) Backgrounder: Efficient Consumer Response (ECR).
http://www.fmi.org/media/bg/ecr1.html [2000-04-27; as of 2001-05-15]

Forrester, J.W. (1958) Industrial Dynamics - a major breakthrough for decision makers. Harvard Business Review 36/4, pp. 37-66

Forrester, J.W. (1961) Industrial Dynamics. MIT Press, Cambridge

Foster, T. (1999) 4PLs: The next generation for supply chain outsourcing? Logistics Management and Distribution Report 38/4.
http://www.manufacturing.net/magazine/logistic/archives/1999/log0401.99/306.htm
[1999-04-01; as of 2001-05-15]

Gebauer, J., Schad, H. (1998) Building an internet-based workflow system - The case of Lawrence Livermore National Laboratories' Zephyr Project. Fisher Center Working Paper 98-WP-1030.
http://haas.berkeley.edu/citm/wp-1030.pdf [1998-04-09; as of 2001-05-15]

Gebauer, J., Segev, A. (2000) Emerging technologies to support indirect procurement: Two case studies from the Petroleum Industry. Information Technology and Management Journal 1/1-2, pp. 107-128.
http://haas.berkeley.edu/~citm/procurement/publications/JITM_final.pdf
[1999-06-10; as of 2001-05-15]

Geoffrion, A.M., Powers, RF (1995) Twenty years of strategic distribution system design: An Evolutionary Perspective. Interfaces 25/5, pp. 105-127

Gilbert, A. (2000) E-Procurement: Problems behind the promise. Information Week 813.
http://www.informationweek.com/813/preprocure.htm [2000-10-20; as of 2001-05-15]

Goetschalckx, M. (2000): Strategic network planning, in: Stadtler, H., Kilger, C. (Eds.) Supply Chain Management and Advanced Planning: Concepts, Models, Software and Case Studies. Springer, Berlin et al., pp. 79-95

Goldratt, E. M. (1990) What is the thing called Theory of Constraints and how should it be implemented? North River Press, Croton-on-Hudson et al.

Grünewald, C. (2000) SAP-APO Projekt bei Aventis. In: Lawrenz, O., Hildebrand, K., Nenninger, M. (Eds.) Supply Chain Management. Vieweg, Braunschweig/Wiesbaden, pp. 281-287

Handfield, R.B., Nichols, Jr., E.L. (1999) Introduction to supply chain management. Prentice Hall, Upper Saddle River

Hantusch, T., Matzke, B., Pérez, M. (1997) SAP R/3 im Internet. Globale Plattform für Handel, Vertrieb und Informationsmanagement. Addison-Wesley, Bonn et al.

Harland, C.M. (1996) Supply chain management: Relationships, chains and networks. British Journal of Management 7/Special Issue, pp. S63-S80

Hax, A.C., Meal, H.C. (1975) Hierarchical integration of production planning and scheduling. In: Geisler, M.A. (Ed.) Logistics. North-Holland/American Elsevier, Amsterdam et al.

Heitmann, S. (2001) Constraint Management - damit die Krise nicht Normalfall wird. Frankfurter Allgemeine Zeitung no vol./67 (2001-03-20), CeBIT Special Edition, p. B23

Herbst, H. (1997) Business rule-oriented conceptual modeling. Physica, Heidelberg

Hicks, D.A. (1997) The manager's guide to supply chain and logistics problem-solving tools and techniques, Part II: Tools, companies, and industries. IIE Solutions 29/10, pp. 24-29

Holt, C.C., Modigliani, F., Shelton, J.P. (1968) The transmission of demand fluctuations through a distribution and production system: The TV-Set industry. Canadian Journal of Economics 1/4, pp. 718-739

Horváth, P. (1991) Schnittstellenüberwindung durch das Controlling. In: Horváth, P. (Ed.) Synergien durch Schnittstellen-Controlling. Poeschel, Stuttgart, pp. 1-23

Houlihan, J. B. (1985) International supply chain management. International Journal of Physical Distribution & Materials Management 15/1, pp. 22-38 (reprinted in: Christopher, M. (Ed.) (1992) Logistics: The Strategic Issues. Chapman&Hall, London, pp. 140-159)

Huang, G.Q., Huang, J., Mak, K.L. (2000) Early supplier involvement in new product development on the internet: Implementation perspectives. Concurrent Engineering: Research and Applications 8/1, pp. 40-49

Huang, G.Q., Mak, K.L. (2000) WeBid: A web-based framework to support early supplier involvement in new product development. Robotics and Computer Integrated Manufacturing 16/2-3, pp. 169-179

Hughes, J., Ralf, M., Michels, B. (1998) Transforming your supply chain. Releasing value in business. International Thompson Business Press, London

Hummeltenberg, W. (1981) Optimierungsmethoden zur betrieblichen Standortwahl. Modelle und ihre Berechnung. Physica, Würzburg et al.

Hurtmanns, F., Packowski, J. (1999) Supply-Chain-Optimierung mit SAP APO in der Chemieindustrie: Einsatzuntersuchung und Geschäftsprozeßszenarien. HMD - Praxis der Wirtschaftsinformatik 36/207, pp. 58-69

Johnson, M.E., Davis, T. (1998) Improving supply chain performance by using order fulfillment metrics. National Productivity Review 17/3, pp. 3-16

Jones, G., Roberts, M. (1990) Optimised Production Technology (OPT). IFS, Kempston

Jones, T.C., Riley, D.W. (1985) Using inventory for competitive advantage through supply-chain management. International Journal for Physical Distribution & Materials Management 15/5, pp. 16-26 (reprinted in: Christopher, M. (Ed.) (1992) Logistics: The Strategic Issues. Chapman&Hall, London, pp. 87-100)

Keebler, J.S., Manrodt, K.B., Durtsche, D.A., Ledyard, D.M. (1999) Keeping SCORE, Measuring the business value of logistics in the supply chain. Council of Logistics Management, Oak Brook

Knolmayer, G. (1987a) Materialflußorientierung statt Materialbestandsoptimierung: Ein Paradigmawechsel in der Theorie des Produktions-Managements? In: Baetge, J., Rühle von Lilienstern, H. Schäfer, H. (Eds.) Logistik - eine Aufgabe der Unternehmenspolitik. Duncker & Humblot, Berlin, pp. 53-77

Knolmayer, G. (1987b) The performance of lot sizing heuristics in the case of sparse demand patterns. In: Kusiak, A. (Ed.) Modern Production Management Systems. North-Holland, Amsterdam et al., pp. 265-280

Knolmayer, G. (2001a) On the optimal extent of mass customization. In: Sebaaly, M.F. (Ed.) Proceedings of the International NAISO Conference on Information Science Innovations. ICSC Academic Press, Slierecht/Millet, pp. 122-128

Knolmayer, G. (2001b) Advanced Planning and Scheduling Systems: Optimierungsmethoden als Entscheidungskriterium für die Beschaffung von Software-Paketen? In: Wagner, U. (Ed.) Zum Erkenntnisstand der Betriebswirtschaftslehre am Beginn des 21. Jahrhunderts. Duncker & Humblot, Berlin, pp. 135-155

Knolmayer, G., Walser, K. (2000) Informationen zum Supply Chain Management im Internet. Wirtschaftsinformatik 42 (2000) 4, pp. 359-370

Kobler, R.-A. (1997) Strategic European distribution logistics design. Dissertation 2059, University of St. Gall. Difo-Druck, Bamberg

Kokkinaki, A.I., Dekker, R., van Nunen, J., Pappis, C. (2000) An exploratory study on electronic commerce for reverse logistics. Supply Chain Forum 1/1, pp. 10-17

Kotzab, H., Otto, A. (2001) Transferring end-user orientation to physical distribution action - considering supply chain management as a logistical marketing approach. http://www.dcpress.com/jmb/kotzab.htm [2001-03-01; as of 2001-05-15]

Kuhn, H., Werner, G. (2000) Supply chain simulation. Paper presented at the Symposium on OR 2000, Dresden

Kurbel, K. (1992) Multimedia-Unterstützung für die Fertigungssteuerung. Zeitschrift für wirtschaftliche Fertigung und Automatisierung 87/12, pp. 664-668

Kurbel, K. (1993) Production scheduling in a leitstand system using a neural-net approach. In: Balagurusamy, E., Sushila, B. (Eds.) Artificial Intelligence Technology. Applications and Management. Tata McGraw-Hill, New Delhi et al., pp. 297-305

Kurbel, K., Loutchko, I. (2001) A framework for mulit-agent electronic marketplaces: Analysis and classificaiton of existing systems. In: Sebaaly, M.F. (Ed.) Proceedings of the International NAISO Conference on Information Science Innovations. ICSC Academic Press, Slierecht/Millet, pp. 335-339

Kurbel, K., Rautenstrauch, C. (1997) Integration des Produktrecycling in die Produktionsplanung und -steuerung. In: Weber, J. (Ed.) Umweltmanagement. Aspekte einer umweltbezogenen Unternehmensführung. Schäffer-Poeschel, Stuttgart, pp. 299-320

Kurbel, K., Zyadeh, H., Schneider, B. (1996) Funktionen, Aufbau und Einsatzformen eines betrieblichen Recycling-Informationssystems. Industrie Management 12/5, pp. 55-60

Kyratsis, E.P., Manson-Partridge, B.M. (1999) Implementing concurrent engineering in supply chain companies and the role of CAE technology. Engineering Designer 25/5, pp. 4-8

Lambert, D.M., Cooper, M.C., Pagh, J.D. (1998) Supply chain management: Implementation issues and research opportunities. The International Journal of Logistics Management 9/2, pp. 1-19

Langan, R., Davis, M., O'Sullivan, D. (1996) Supply chain management: A case study of AsIs vs. ToBe. In: Browne, J. et al. (Eds.) IT and Manufacturing Partnerships, IOS Press, Amsterdam

Lapide, L. (1998): Supply chain planning optimization: Just the facts. http://www.amrresearch.com/preview/9805scsexec.asp [1998-05; as of 2001-05-15]

Lee, H.L. (1993) Design for supply chain management: Concepts and examples. In: Sarin, R.K. (Ed.) Perspectives in Operations Management. Kluwer, Boston et al., pp. 45-65

Lee, H.L. (1998) Postponement for mass customization. In: Gattorna, J. (Ed.) Strategic Supply Chain Alignment. Best practice in supply chain management. Gower, Brookfield, pp. 77-91

Lee, H.L., Billington, C. (1992) Managing supply chain inventory: Pitfalls and opportunities. Sloan Management Review 33/3, pp. 65-73

Lee, H.L., Padmanabhan, V., Whang, S. (1997a) Information distortion in a supply chain: The bullwhip effect. Management Science 43/4, pp. 546-558

Lee, H.L., Padmanabhan, V., Whang, S. (1997b) The bullwhip effect in supply chains. Sloan Management Review 38/3, pp. 93-102

Lee, H., Whang, S. (1999) Decentralized multi-echelon supply chains: Incentives and information. Management Science 45/5, pp. 633-640

Le Pape, C. (1994) Implementation of resource constraints in ILOG SCHEDULE: A library for the development of constraint-based scheduling systems. Intelligent Systems Engineering 3/2, pp. 55-66

Lin, G., Ettl, M., Buckley, S., Bagchi, S., Yao, D.D., Naccarato, B.L., Allan, R., Kim, K., Koenig, L. (2000) Extended-enterprise supply-chain management at IBM Personal Systems Group and other divisions. Interfaces 30/1, pp. 7-25

Mak, K.L., Wong, Y.S. (1997) A genetic approach to Kanban assignment problems. Journal of Information & Optimization Sciences 18/3, pp. 359-381

Makridakis, S., Wheelwright, S.C., Hyndman, R.J. (1997) Forecasting: Methods and applications. 3rd ed., Wiley, New York et al.

Martin, M., Hausman, W., Ishii, K. (1998) Design for variety. In: Ho, T.-H., Tang, C.S. (Eds.) Product Variety Management. Kluwer, Boston et al., pp. 103-122

Mason-Jones, R., Towill, D.R. (1997) Information enrichment: designing the supply chain for competitive advantage. Supply Chain Management 2/4, pp. 137-148

Mason-Jones, R., Towill, D.R. (2000) Coping with uncertainty: Reducing "Bullwhip" behaviour in global supply chains. Supply Chain Forum 1/1, pp. 40-45

McMullen, Jr., T.B. (1998) Introduction to the Theory of Constraints (TOC) management system. St. Lucie Press, Boca Raton et al.

Mertens, P. (Ed.) (1994a) Prognoserechnung. 5th ed., Physica, Heidelberg

Mertens, P. (1994b) Neuere Entwicklungen des Mensch-Computer-Dialoges in Berichts- und Beratungssystemen. Zeitschrift für Betriebswirtschaft 64/1, pp. 35-55

Mertens, P. (1995) Supply Chain Management (SCM). Wirtschaftsinformatik 37/2, pp. 177-179

Mertens, P. (2000) Integrierte Informationsverarbeitung 1. Administrations- und Dispositionssysteme in der Industrie. 12th ed., Gabler, Wiesbaden

Mertens, P., Knolmayer, G. (1998) Organisation der Informationsverarbeitung. Grundlagen - Aufbau - Arbeitsteilung. 3rd ed., Gabler, Wiesbaden

Mertens, P., Zeier, A. (1999) ATP - Available-to-Promise. Wirtschaftsinformatik 41/4, pp. 378-379

Metters, R. (1997) Quantifying the bullwhip effect in supply chains. Journal of Operations Management 15/1, pp. 89-100

Meyer, H., Rohde, J., Wagner, M. (2000) Architecture of selected APS. In: Stadtler, H., Kilger, C. (Eds.) Supply Chain Management and Advanced Planning. Springer, Berlin et al., pp. 241-249

Monczka, R., Trent, R., Handfield, R. (1998) Purchasing and supply chain management. South-Western College Publishing, Cincinnati

Monden, Y. (1983) Toyota production system. Institute of Industrial Engineering, Norcross

Moser, G. (1999) SAP R/3 interfacing using BAPIs: A practical guide to working within the SAP Business Framework. Vieweg, Braunschweig et al.

MSI (2001): Glossary.
http://www.manufacturingsystems.com/glossary/ [2001; as of 2001-05-15]

New, S.J. (1998) Simple systems - simple models? In: Bititci, U.S., Carrie, A.S. (Eds.) Strategic Management of the Manufacturing Value Chain. Kluwer, Boston et al., pp. 553-560

Nicholas, J.M. (1998) Competitive manufacturing management. Continuous improvement, lean production. Customer-Focused Quality. Irwin/McGraw-Hill, Boston et al.

NN (1998) Distribution-intensive supply chains growing with APS market. IIE Solutions 30/6, p. 6

NN (1999a) The Virtual Buyer.
http://purchserv.finop.umn.edu/purch/phelp.html [1999-12-20; as of 2001-05-15]

NN (1999b) OBI Standard continues to gain acceptance.
http://www.openbuy.org/obi/news/press/1999/011299.html [1999-01-12; as of 2001-05-15]

NN (1999c) Ariba Supports Commerce XML (cXML).
http://www.oasis-open.org/cover/aribaCXML.html [1999-02-08; 2000-12-27; as of 2001-05-15]

NN (1999d) Carriers and 3PLs embrace ERP software. Logistics Management & Distribution Report 38/4.
http://www.manufacturing.net/magazine/logistic/archives/1999/log0401.99/296.htm [1999-01-04; as of 2001-05-15]

NN (1999e) How PRECOR stays fit (Reprinted from Start Magazine, 1999-09). http://www.agilesoft.com/reviews/start2.asp [1999-09; as of 2001-05-15]

NN (1999f) SAP blockiert den Markt für Logistiksoftware. Computer Zeitung 30/29, p. 23

NN (1999g) Power of APS optimisation engine is key. http://www.insight-logistics.co.uk/press_releases/gig_engine_5_99.htm [1999-05; as of 2001-05-15]

NN (2000) Private profiles: eTime capital. ComputerLetter 16/33. http://www.etimecapital.com/news_events/news_events_artcl_compltr_100200.htm [2000-10-02; as of 2001-05-15]

NN (n. d.) What is ELEMICA? http://www.elemica.com/elemdotcom/new_faqs.jsp?bmUID=989938556383 [no date; as of 2001-05-15]

Norris, G., Hurley, J.R., Hartley, K.M., Dunleavy, J.R., Balls, J.D. (2000) E-Business and ERP - Transforming the enterprise. Wiley, New York et al.

Oertel, H.A., Abraham, H.-W. (1994) Prozeßgestaltung durch Einkauf und Logistik. In: Koppelmann, U., Lumbe, H.-J. (Eds.) Prozeßorientierte Beschaffung. Schäffer-Poeschel, Stuttgart, pp. 171-191

Oliver, R.K., Webber, M.D. (1982) Supply-chain management: logistics catches up with strategy. Outlook (1982) (reprinted in: Christopher, M. (Ed.) (1992) Logistics: The Strategic Issues. Chapman&Hall, London, pp. 63-75)

Orlicky, J. (1975) Material requirements planning. The new way of life in production and inventory management. McGraw-Hill, New York et al.

Parsons, M.G., Singer, D.J., Sauter, J.A. (1999) A hybrid agent approach for set-based conceptual ship design. International Conference on Computer Applications in Shipbuilding, Cambridge. http://www.erim.org/cec/papers/AHybridAgent.pdf [1999-06-07; as of 2001-05-15]

Pfohl, H.-C. (2000) Logistiksysteme. Betriebswirtschaftliche Grundlagen. 6th ed., Springer, Berlin et al.

Phelps, T. (1997) AutoSTEP business process activities. Autotech 1997. http://www.erim.org/~vparunak/scdocs/autostep/ [1997-09-15; 1998-09-09; as of 2001-05-15]

Philippson, C., Pillep, R., von Wrede, P., Röder (1999) Marktspiegel Supply Chain Management Software. Forschungsinstitut für Rationalisierung, Aachen

Pinedo, M. (1995) Scheduling - Theory, algorithms, and systems. Prentice-Hall, Englewood Cliffs

Poirier, C.C. (1999) Advanced supply chain management: How to build a sustained competitive advantage. Berrett-Koehler, San Francisco

Porter, A.M. (1998) UTC puts supply managers through tough basic training. Purchasing online. http://www.manufacturing.net/magazine/purchasing/archives/1998/pur0423.98/042news.htm [1998-04-23; as of 2001-05-15]

PriceWaterhouseCoopers (Ed.) (1999) Information and technology in the supply chain. Making technology pay. Euromoney Publications, London

Ptak, C.A., Schragenheim, E. (2000) ERP - Tools, techniques, and applications for integrating the supply chain. St. Lucie Press/APICS, Boca Raton et al./Falls Church

Rautenstrauch, C. (1997) Fachkonzept für ein integriertes Produktions-, Recyclingplanungs- und Steuerungssystem (PRPS-System). de Gruyter, Berlin et al.

Razzaque, M.A., Sheng, C.C. (1998) Outsourcing of logistics functions: A literature survey. International Journal of Physical Distribution & Logistics Management 28/2, pp. 89-107

Rich, N., Hines, P. (1997) Supply-Chain management and time-based competition: The role of the supplier association. International Journal of Physical Distribution & Logistics Management 27/3, pp. 210-225

Riggs, D.A., Robbins, S.L. (1998) The executive's guide to supply management strategies: Building supply chain thinking into all business processes. Amacom, New York et al.

Rodens-Friedrich, B. (1999) ECR bei dm-drogerie markt - Unser Weg in die Wertschöpfungspartnerschaft. In: von der Heydt, A. (Ed.) Handbuch Efficient Consumer Response: Konzepte, Erfahrungen, Herausforderungen. Vahlen, München, pp. 205-221

Ross, D.F. (1999) Competing through supply chain management. Creating market-winning strategies through supply chain partnerships. Kluwer, Boston et al.

Sakakibara, S., Flynn, B.B., Schroeder, R.G. (1993) A framework and measurement instrument for just-in-time manufacturing. Production and Operations Management 2/3, pp. 177-194

Saunders, M. (1997) Strategic purchasing and supply chain management. 2nd ed., Financial Times Management, London et al.

Schaarschmidt, R., Röder, W. (1997) Datenbankbasiertes Archivieren im SAP System R/3. Wirtschaftsinformatik 39/5, pp. 469-477

Schäffer, H., Höll, S., Schönberg, T. (1999) Buy Direct - Eine Internet-basierende Geschäftsprozeßoptimierung im Einkauf. In: Scheer, A.-W., Nüttgens, M. (Eds.) Electronic Business Engineering. Physica, Heidelberg, pp. 505-519

Schär, M. (1999) Revolution im Liftschacht. Schindlers neuer, maschinenraumloser Lift spart viel Platz und viel Geld. Cash 11/10, p. 15

Schaumann, U. (1991) Neugestaltung der logistischen Kette in der Produktion nach der Just-in-Time-Philosophie. In: Rupper, P. (Ed.) Unternehmenslogistik. 3rd ed., Industrielle Organisation/TÜV Rheinland, Zürich, pp. 115-134

Scheer, A.-W. (1994) CIM - Computer Integrated Manufacturing. Towards the Factory of the Future, 3rd ed., Springer, Berlin et al.

Scheuerer, A. (1995) Beiträge zur Steuerung des betrieblichen Recyclings unter besonderer Berücksichtigung eines Informationssystems zur Unterstützung von Demontageprozessen. Dissertation University of Erlangen-Nürnberg

Schmidt, C., Weinhardt, C., Horstmann, R. (1998) Internet-Auktionen. Eine Übersicht zu Online-Versteigerungen im Hard- und Softwarebereich. Wirtschaftsinformatik 40/5, pp. 450-458

Schneider, R., Grünewald, C. (2000) Supply Chain Management-Lösung mit mySAP.com. In: Lawrenz, O., Hildebrand, K., Nenninger, M. (Eds.) Supply Chain Management. Vieweg, Braunschweig/Wiesbaden, pp. 145-177

Schönsleben, P. (2000) Integral Logistics Management, Planning & Control of Comprehensive Business Processes. St. Lucie Press/APICS, Boca Raton et al./Alexandria

Schulte, U.G., Hoppe, T. (1999) Ausrichtung der Versorgungskette auf den Konsumenten – Supply Chain Management bei Elida Fabergé. In: von der Heydt, A. (Ed.) Handbuch Efficient Consumer Response: Konzepte, Erfahrungen, Herausforderungen. Vahlen, München, pp. 65-75

SCN Education (Eds.) (2000) ASP - Application Service Providing. Vieweg/Gabler, Braunschweig/Wiesbaden

Seal, W., Cullent, J., Dunlop, A., Berry, T., Ahmed, M. (1999) Enacting a European supply chain: a case study on the role of management accounting. Management Accounting Research 10/3, pp. 303-322

Segev, A., Gebauer, J. (1998) Procurement transformation through web-based technologies. http://www.haas.berkeley.edu/citm/wip/924-3Segev_Gebauer.pdf [1998-09-24; as of 2001-05-15]

Seifert, W. (1999) Gestaltung von Logistikzentren im stationären Handel unter Berücksichtigung aktueller ECR-Logistikstrategien und verändertem Kundenverhalten. In: von der Heydt, A. (Ed.) Handbuch Efficient Consumer Response: Konzepte, Erfahrungen, Herausforderungen. Vahlen, München, pp. 88-96

Sellers, P. (1992) The dumbest marketing ploy. Fortune 126/5, pp. 88-93

Shams, K. (2000): Know ThySAP.COM. http://www.intelligenterp.com/feature/2000/10/shams.shtml [2000-10-20; as of 2001-05-15]

Siemieniuch, C.E., Sinclair, M.A. (2000) Implications of the supply chain for role definitions in concurrent engineering. Human Factors and Ergonomics in Manufacturing 10/3, pp. 251-272

Singh, M.P. (1999) The end of the supply chain? IEEE Internet Computing 3/6, pp. 4-5

Slats, P.A., Bhola, B., Evers, J.J.M., Dijkhuizen, G. (1995) Logistic chain modeling. European Journal of Operational Research 87/1, pp. 1-20

Smith, B. (1999) The future of supply chain management on the internet. Supply Chain & Logistics Journal 1/4.
http://www.infochain.org/quarterly/Feb99/Smith.html [1999-02; 2000-05-25; as of 2001-05-15]

Snodgrass, R.T. (Ed.) (1995) The TSQL2 temporal query language. Kluwers, Boston et al.

Sobolewski, M. (1996) Multiagent knowledge-based environment for concurrent engineering applications. Concurrent Engineering: Research and Applications 4/1, pp. 89-97

SoftGuide Softwareführer (2001) CATRIN.
http://www.softguide.de/prog_c/pc_0100.htm [2001-05-09; as of 2001-05-15]

Stadtler, H., Kilger, C. (Eds.) (2000) Supply chain management and advanced planning, Springer, Berlin et al.

Steele, W. (1999), quoted in NN (1999g).
http://www.insight-logistics.co.uk/press_releases/gig_engine_5_99.htm

Stein, T. (1997) Not just ERP anymore. InformationWeek Online.
http://www.iweek.com/659/59iuerp.htm [1997-12-01; as of 2001-05-15]

Stein, T. (1998) Extending - Companies that don't use Enterprise Resource Planning software to share information may regret it. InformationWeek no vol./686.
http://www.techweb.com/se/directlink.cgi?IWK19980615S0044 [1998-06-15; as of 2001-05-15]

Stephens, S. (2000) Supply Chain Council & Supply Chain Operations Reference (SCOR) Model Overview.
http://www.supply-chain.org/Resources/scor_overview.cfm [2000-05; as of 2001-05-15]

Sterman, J.D. (1989) Modeling managerial behavior: Misperceptions of feedback in a dynamic decision making experiment. Management Science 35/3, pp. 321-339

Stock, S. (2001) Modellierung zeitbezogener Daten im Data Warehouse. Gabler/Deutscher Universitäts-Verlag, Wiesbaden

Stürken, M. (2001) Produktdatenmanagementsystem.
http://www.wi1.uni-erlangen.de/buecher/lexikon/pdm.html [2001-04-11; as of 2001-05-15]

Sturrock, J.A. (1999) In-house or third party distribution? In: Hadjiconstantinou, E. (Ed.) Quick Response in the Supply Chain. Springer, Berlin et al., pp. 51-60

Sweat, J. (1998) FedEx and SAP team on tracking-shipping app. InformationWeek Online.
http://www.informationweek.com/677/77iufdx.htm [1998-04-13; as of 2001-05-15]

Szuprowicz, B.O. (2000) Supply chain management for e-business infrastructures. Computer Technology Research, Charleston

Takahashi, K., Nakamura, N. (1999) Reacting JIT ordering systems to the unstable changes in demand. International Journal of Production Research 37/10, pp. 2293-2313

Taylor, D., Brunt, D. (Eds.) (2001) Manufacturing operations and supply chain management: The LEAN approach. Thomson Learning, London

Tayur, S., Ganeshan, R., Magazine, M. (Eds.) (1999) Quantitative models for supply chain management. Kluwer, Boston et al.

Thru-put Technologies (2001) What color is your supply chain? http://www.thru-put.com/technology/supply.html [2001-03-30; as of 2001-05-15]

Tompkins, J.A. (2000) No boundaries: Moving beyond supply chain management. Tompkins Press, Raleigh

Toomey, J.W. (1996) MRP II. Planning for manufacturing excellence. Chapman&Hall, New York et al.

Towill, D.R. (1994) 1961 and all that: The influence of Jay Forrester and John Burbidge on the design of modern manufacturing systems. 1994 International Systems Dynamics Conference on Business Decision-Making, Systems Dynamics Society, Cambridge, pp. 105-115

Towill, D.R. (1996) Industrial dynamics modelling of supply chains. International Journal of Physical Distribution & Logistics Management 26/2, pp. 23-42

Towill, D.R. (1997) FORRIDGE - Principles of good practice in material flow. Production Planning & Control 8/7, pp. 622-632

Towill, D. (1999) Simplicity Wins: Twelve rules for designing effective supply chains. Control 25/2, pp. 9-13

Towill, D.R., Naim, M.M., Wikner, J. (1992) Industrial dynamics simulation models in the design of supply chains. International Journal of Physical Distribution & Logistics Management 22/5, pp. 3-13

Tsay, A.A., Lovejoy, W.S. (1999) Quantity flexibility contracts and supply chain performance. Manufacturing & Service Operations Management 1/2, pp. 89-111

Tyndall, G., Gopal, C., Partsch, W., Kamauff, J. (1998) Supercharging supply chains. New ways to increase value through global operational excellence. Wiley, New York et al.

Ulrich, K., Randall, T., Fisher, M., Reibstein, D. (1998) Managing product variety. In: Ho, T.-H., Tang, C.S. (Eds.) Product Variety Management. Kluwer, Boston et al., pp. 177-205

Underhill, T. (1996) Strategic Alliances. Managing the supply chain. PennWell Publishing, Tulsa

Urban, G. (1998) Anregungen für Logistik-Systeme - aus Sicht eines Nutzfahrzeugherstellers. Information Management & Consulting 13/3, pp. 42-45

Van Hentenryck, P. (1999) The OPL Optimization Programming Language. MIT Press, Cambridge/London

van Hoek, R.I. (1998) Reconfiguring the supply chain to implement postponed manufacturing. The International Journal of Logistics Management 9/1, pp. 95-110

Vaterrodt, J.C. (1995) Recycling zwischen Betrieben. Stand und Perspektiven der zwischenbetrieblichen Rückführung von Produktions- und Konsumtionsrückständen in die Fertigungsprozesse von Unternehmen. Schmidt, Berlin

Vazsonyi, A. (1954) The use of mathematics in production and inventory control. Management Science 1/1, pp. 70-85

Verwijmeren, M., van der Vlist, P., van Donselaar, K. (1996) Networked inventory management information systems: materializing supply chain management. International Journal of Physical Distribution & Logistics 26/6, pp. 16-31

Vidal, C.J., Goetschalckx, M. (1997) Strategic production-distribution models: A critical review with emphasis on global supply chain models. European Journal of Operational Research 98/1, pp. 1-18

Vollmann, T.E., Berry, W.L., Whybark, D.C. (1992) Manufacturing planning and control systems. 3rd ed., Irwin, Homewood

Voss, S., Martello, S., Roucairol, C., Osman, I.H. (Eds.) (1998) Meta-Heuristics: Advances and trends in local search paradigms for optimization. Kluwer, Boston et al.

Voß, S., Woodruff, D.L. (2000) Supply chain planning: Is mrp a good starting point? In: Wildemann, H. (Ed.) Supply Chain Management. Transfer-Centrum-Verlag, München, pp. 177-203

Warnecke, H.-J. (1993) The fractal company: A revolution in corporate culture. Springer, Berlin et al.

Weiß, E. (1996) Optimierung von Produktionsnetzwerken auf der Basis des "Wirtschaftsglobus-Modells". In: Wildemann, H. (Ed.) Produktions- und Zuliefernetzwerke. TCW, München, pp. 105-144

White, R.E., Pearson, J.N., Wilson, J.R. (1999) JIT Manufacturing: A Survey of Implementations in Small and Large U.S. Manufacturers. Management Science 45/1, pp. 1-15

Wiendahl, P. (1995) Load-oriented manufacturing control. Springer, Berlin et al.

Wildemann, H. (1985) Implementation strategies for the integration of Japanese Kanban-principles in German companies. Engineering Costs and Production Economics 9, pp. 305-319

Wildemann, H. (1994) Fertigungsstrategien. Reorganisationskonzepte für eine schlanke Produktion und Zulieferung. 2nd ed., Transfer-Centrum, München

Wilding, R. (1998) Chaos, Complexity and Supply-Chains. Logistics Focus 6/8, pp. 8-10

Williams, H.-P., Wilson, J.M. (1998) Connections between integer linear programming and constraint logic programming - An overview and introduction to the cluster of articles. INFORMS Journal on Computing 10/3, pp. 261-264

Woods, J.A., Marien, E.J. (Eds.) (2001) The supply chain yearbook, 2001 edition. McGraw-Hill, New York et al.

Zäpfel, G., Missbauer, H. (1993) New concepts for production planning and control. European Journal of Operational Research 67/3, pp. 297-320

Zäpfel, G., Piekarz, B. (1996) Supply Chain Controlling. Interaktive und dynamische Regelung der Material- und Warenflüsse. Ueberreuter, Wien

Zeng, D., Sycara, K. (1997) How can an agent learn to negotiate? In: Müller, J.P., Wooldridge, M.J., Jennings, N.R. (Eds.) Intelligent Agents III. Agent Theories, Architectures, and Languages. Springer, Berlin et al., pp. 233-244

Zheng, S., Yen, D.C., Tarn, J.M. (2000) The new spectrum of the cross-enterprise solution: The integration of supply chain management and enterprise resources planning systems. Journal of Computer Information Systems 41/1, pp. 84-93

Index

C

G

H

I

J

K

L

M

N

O

P

Q

R

S

T

U